THE
HANDY
ANATOMY
ANSWER
BOOK

Also from Visible Ink Press

The Handy Answer Book for Kids (and Parents)

The Handy Astronomy Answer Book

The Handy Biology Answer Book

The Handy Geography Answer Book

The Handy Geology Answer Book

The Handy History Answer Book, 2nd Ed.

The Handy Ocean Answer Book

The Handy Math Answer Book

The Handy Physics Answer Book

The Handy Politics Answer Book

The Handy Presidents Answer Book

The Handy Religion Answer Book

The Handy Science Answer Book, Centennial Ed.

The Handy Sports Answer Book

The Handy Supreme Court Answer Book

The Handy Weather Answer Book

Please visit us at visibleink.com.

About the Authors

Naomi E. Balaban, a reference librarian for more than fifteen years at the Carnegie Library of Pittsburgh, has extensive experience in the areas of science and consumer health. She edited, with James Bobick, *The Handy Science Answer Book* and *The Handy Biology Answer Book*. She has a background in linguistics and a master's degree in library science.

James E. Bobick recently retired after sixteen years as Head of the Science and Technology Department at the Carnegie Library of Pittsburgh. During the same time, he taught the science resources course in the School of Information Sciences at the University of Pittsburgh. He has master's degrees in both biology and library science.

THE
HANDY
ANATOMY
ANSWER
BOOK

NAOMI E. BALABAN AND JAMES E. BOBICK

VISIBLE
INK
PRESS

Detroit

THE HANDY ANATOMY ANSWER BOOK

Visible Ink Press®
43311 Joy Rd., #414
Canton, MI 48187-2075

Visible Ink Press is a registered trademark of Visible Ink Press LLC.

Most Visible Ink Press books are available at special quantity discounts when purchased in bulk by corporations, organizations, or groups. Customized printings, special imprints, messages, and excerpts can be produced to meet your needs. For more information, contact Special Markets Director, Visible Ink Press, www.visibleink.com, or 734-667-3211.

Managing Editor: Kevin S. Hile
Art Director: Mary Claire Krzewinski
Typesetting: Marco Di Vita
ISBN 978-1-57859-574-7
ISBN 978-1-57859-500-6

Cover images:
Cell graphic courtesy of Cohen B.J., and Wood, D.L. *Memmler's The Human Body in Health and Disease.* 9th Ed. Philadelphia: Lippincott, Williams & Wilkins, 2000.

Brain illustration courtesy of Cohen, B.J. *Medical Terminology.* 4th Ed. Philadelphia: Lippincott, Williams & Wilkins, 2003.
All other images courtesy of *iStockphoto.com*

Library of Congress Cataloging–in–Publication Data
Balaban, Naomi E.
  The handy anatomy answer book / Naomi E. Balaban and James E. Bobick
    p. ; cm.
  Includes bibliographical references and index.
  ISBN 978-1-57859-190-9 (alk. paper)
    1. Human anatomy—Miscellanea. 2. Physiology—Miscellanea. I. Balaban, Naomi E. II. Title.
    [DNLM: 1. Anatomy—Examination Questions. 2. Anatomy. 3. Physiology—Examination Questions. 4. Physiology. QS 4 B663h 2008]
    QM23.2.B62 2008
    611.0076—dc22

                                                          2008001595

Printed in China.

Contents

Foreword

Pick up a magazine or newspaper, turn on the radio or television, or access medical and health information on the Internet and you will find that the human body is definitely in the news. Artificial hearts, dietary supplements, stem cell research, genetic engineering, arthroscopic surgery, and many other intriguing subjects on human biology and health are talked about daily. Our bodies seem to be constantly on our minds! *The Handy Anatomy Answer Book* is here to help you unravel the complexities and mysteries of how your body works.

Our interest in and understanding of the human body has a long and detailed history going back to the ancient Greeks Aristotle and Galen, who first studied the structure and function of our species. From this starting point, however, the scientific study of the body progressed slowly. It wasn't until the sixteenth century that the foundations of modern anatomy were laid by Andreas Vesalius; and it wasn't until the century after that when William Harvey discovered how blood circulates in the body. Finally, in the nineteenth century, anatomy and physiology became distinct scientific disciplines.

As techniques for making more accurate observations and performing careful experiments evolved, knowledge of the human body expanded rapidly. And as our knowledge expanded, so did the vocabulary to describe what medical doctors were discovering. Using the root languages of Greek and Latin, soon there was a plethora of complex terms describing body parts, their precise locations, and their functions.

The Handy Anatomy Answer Book helps make the language of anatomy—as well as physiology and pathology—more understandable and less intimidating to the general reader, while answering over one thousand questions on all the major body systems.

In this informative book, you'll find answers to such intriguing questions as: Who discovered how muscles work? What is the largest nerve in the body? How much air can your lungs hold? What are the primary sensations of taste? Who is considered the founder of physiology? How many bones are in the human body? *The Handy Anatomy Answer Book* also includes fascinating trivia. Do identical twins share the same fingerprints? What tissues in your body can regenerate? Does brain size affect intelligence?

We are pleased and excited to contribute another addition to the "Handy Answer" family. It is wonderful to be a part of the continued growth of this series, which began in 1994 with *The Handy Anatomy Answer Book*.

—Naomi E. Balaban and James E. Bobick

Acknowledgments

The authors thank their families for their ongoing interest, encouragement, support, and especially their understanding while this volume was taking shape and being transformed from a proposal to the finished product. In particular, Naomi thanks Carey for his patience and expertise; and, Jim thanks Sandi, especially for her extraordinary assistance.

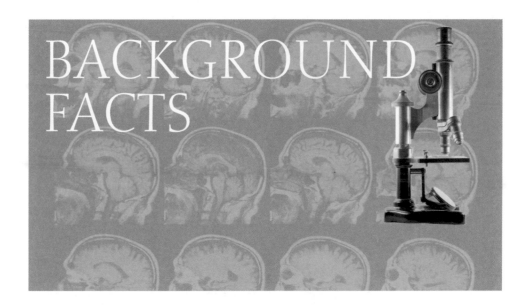

BACKGROUND FACTS

HISTORY

Which **scientific disciplines** study the **human body**?

The scientific disciplines of anatomy and physiology study the human body. Anatomy (from the Greek *ana* and *temnein,* meaning "to cut up") is the study of the structure of the body parts, including their form and organization. Physiology (from the Latin, meaning "the study of nature") is the study of the function of the various body parts and organs. Anatomy and physiology are usually studied together to achieve a complete understanding of the human body.

How is the field of **anatomy** divided into **subdivisions**?

The field of anatomy is generally divided into macroscopic, or gross anatomy (not requiring a microscope), and microscopic anatomy. Gross anatomy includes the subdivisions of regional anatomy, systemic anatomy, developmental anatomy, and clinical anatomy. Regional anatomy studies specific regions of the body, such as the head and neck or lower and upper limbs. Systemic anatomy studies different body systems, such as the digestive system and reproductive system. Developmental anatomy describes the changes that occur from conception through physical maturity. Clinical anatomy includes medical anatomy (anatomical features that change during illness) and radiographic anatomy (anatomical structures seen using various imaging techniques).

The two major subdivisions of microscopic anatomy are cytology and histology. Cytology (from the Greek *cyto,* meaning "cell") is the study and analysis of the internal structure of individual cells. Histology (from the Greek *histos,* meaning "web") is the study and examination of tissues.

Most people consider Hippocrates the founder of the discipline of medicine; and the Hippocratic oath—the code of ethics followed by physicians—is also named after him. © iStockphoto.com/Phil Sigin.

What are some **specialties** of **physiology**?

Specialties and subdivisions of physiology include cell physiology, special physiology, systemic physiology, and pathological physiology, often called simply pathology. Cell physiology is the study of the functions of cells, including both chemical processes within cells and chemical interactions between cells. Special physiology is the physiological study of specific organs, such as cardiac physiology, which is the study of heart function. Systemic physiology is comparable to systemic anatomy since it is the study of the functions of different body systems, such as renal physiology and neurophysiology. Pathology (from the Greek *pathos,* meaning "suffering" or "disease") is the study of the effects of diseases on organs or systems and diseased cells and tissues.

When did the study of **anatomy and physiology** first become **accepted** as **sciences**?

Anatomy and physiology were first accepted as sciences during ancient Greek times. Hippocrates (c. 460–377 B.C.E.), who is considered the father of medicine, established medicine as a science, separating it from religion and philosophy. His application of logic and reason to medicine was the beginning of observational medicine.

What were **Aristotle's contributions** to **anatomy**?

Aristotle (384–322 B.C.E.) wrote several works laying the foundations for comparative anatomy, taxonomy, and embryology. He investigated carefully all kinds of animals, including humans. His works on life sciences, *On Sense and Sensible Objects, On Memory and Recollection, On Sleep and Waking, On Dreams, On Divination by Dreams, On Length and Shortness of Life, On Youth and Age,* and *On Respiration,* are collectively called *Parva Naturalia.*

Who is considered the **father of physiology**?

The Greek physician and anatomist Erasistratus (304–250 B.C.E.) is considered the father of physiology. Based on his numerous dissections of human cadavers, he accurately described the brain, including its cavities and membranes, stomach muscles, and the differences between motor and sensory nerves. He understood correctly that the heart served as a pump to circulate blood. Anatomical research ended with Erasistratus until the thirteenth century, in a large part because of public opinion against the dissection of human cadavers.

Whose work during the **Roman era** became the **authority on anatomy**?

Galen (130–200), a Greek physician, anatomist, and physiologist living during the time of the Roman Empire, was one of the most influential and authoritative authors on medical subjects. His writings include *On Anatomical Procedures, On the Usefulness of the Parts of the Body, On the Natural Faculties,* and hundreds of other treatises. Since human dissection was forbidden, Galen made most of his observations on different animals. He correctly described bones and muscles and observed muscles working in contracting pairs. He was also able to describe heart valves and structural differences between arteries and veins. While his work contained many errors, he provided many accurate anatomical details that are still regarded as classics. Galen's writings were the accepted standard text for anatomical studies for 1,400 years.

The Greek philosopher Aristotle helped lay the foundations of several scientific fields, including anatomy. © iStockphoto.com/Phil Sigin.

Who became known as the "**reformer of anatomy**" during the **Renaissance**?

Andreas Vesalius (1514–1564) became known as the "reformer of anatomy" during the Renaissance. His masterpiece and most famous work, *De Humani Corporis Fabrica,* published in 1543, described various body systems and individual organs. It also included beautiful anatomical illustrations. Vesalius challenged many of Galen's teachings, which had become accepted as fact though they were incorrect.

Who improved the **microscope** in a way that greatly impacted **anatomy** and **physiology** studies?

Anton van Leeuwenhoek (1632–1723) was a Dutch microscopist and scientist. Although he did not invent the microscope, he greatly improved the capability of the microscope. His expert skill in grinding lenses achieved a magnification of 270 times, which was far greater than any other microscope of the era. He was able to observe bacteria, striations in muscle, blood cells, and spermatozoa.

What **discovery** of the 17th century helped **establish** the science of **physiology**?

The English physician William Harvey (1578–1657) published *On the Movement of the Heart and Blood in Animals* in 1628. This important medical treatise proved

3

that blood continuously circulated within the vessels. Harvey's discoveries contradicted many beliefs about blood circulation that dated back to the time of Galen. Harvey is considered the father of modern physiology for introducing the experimental method of scientific research.

Who is considered the **founder** of **experimental medicine** and **physiology**?

The French physiologist Claude Bernard (1813–1878) is credited with originating the experimental approach to medicine and establishing general physiology as a distinct discipline. His classic work, *Introduction to the Study of Experimental Medicine,* was published in 1865. He was elected to the Académie Française in 1869 for this work.

The development of the modern microscope vastly improved scientists' knowledge of bacteria, viruses, cells, and small anatomical structures. © iStockphoto.com/Christopher Pattberg Fotodesign.

What was the **first professional organization** of **physiologists**?

The first organization of physiologists was the Physiological Society founded in 1876 in England. In 1878 the *Journal of Physiology* began publication as the first journal dedicated to reporting results of research in physiology. The American counterpart, the American Physiological Society, was founded in 1887. The American Physiological Society first sponsored publication of the *American Journal of Physiology* in 1898.

LEVELS OF ORGANIZATION

What are the **levels** of **structural organization** in vertebrate animals, including humans?

Every vertebrate animal has four major levels of hierarchical organization: cell, tissue, organ, and organ system. Each level in the hierarchy is of increasing complexity, and all organ systems work together to maintain life.

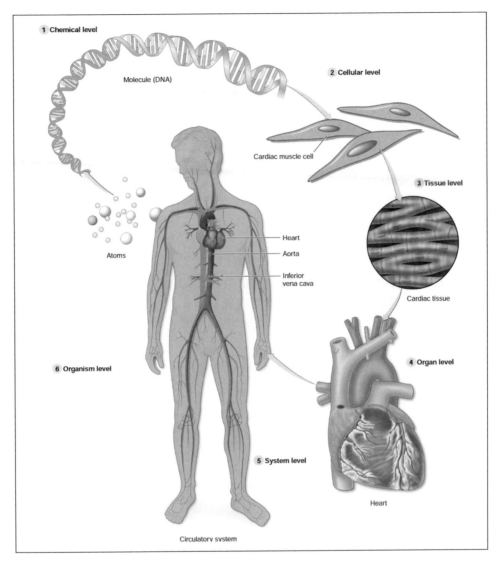

Cells in the human body organize themselves into increasingly complex structures and systems. (From Premkumar K. *The Massage Connection Anatomy and Physiology.* Baltimore: Lippincott, Williams & Wilkins, 2004.)

What is a **cell**?

A cell is a membrane-bound unit that contains hereditary material (DNA) and cytoplasm; it is the basic structural and functional unit of life. See chapter 2 for more details.

What are the **four major types** of **tissue**?

A tissue (from the Latin *texere,*, meaning "to weave") is a group of similar cells that perform a specific function. The four major types of tissue are epithelial, connective, muscle, and nerve. Each type of tissue performs different functions.

5

What is the cell theory?

The cell theory states that the cell is the fundamental component of all life and all organisms are made up of cells. There are three basic principles to the cell theory. First, the cell is the simplest collection of matter that can live. There are diverse forms of life existing as single-celled organisms. More complex organisms, including plants and animals, are multicellular cooperatives composed of diverse, specialized cells that could not survive for long on their own. Secondly, all cells come from preexisting cells and are related by division to earlier cells that have been modified in various ways during the long evolutionary history of life on earth. Finally, all of the life processes of an organism occur fundamentally at the cellular level.

What are the **general characteristics** of the different types of **tissue**?

Each of the four major types of tissue have different functions, are located in different parts of the body, and have certain distinguishing features. The table below explains these differences.

Characteristics of Tissues

Tissue	Function	Location	Distinguishing Features
Epithelial	Protection, secretion, absorption, excretion	Covers body surfaces, covers and lines internal organs, compose glands	Lacks blood vessels
Connective	Bind, support, protect, fill spaces, store fat, produce blood cells	Widely distributed throughout the body	Matrix between cells, good blood supply
Muscle	Movement	Attached to bones, in the walls of hollow internal organs, heart	Contractile
Nervous	Transmit impulses for coordination, regulation, integration, and sensory reception	Brain, spinal cord, nerves	Cells connect to each other and other body parts

What is an **organ**?

An organ is a group of several different tissues working together as a unit to perform a specific function or functions. Each organ performs functions that none of the component tissues can perform alone. This cooperative interaction of different tissues is a basic feature of animals, including humans. The heart is an example of an organ. It consists of cardiac muscle wrapped in connective tissue. The heart chambers are lined with epithelium. Nerve tissue controls the rhythmic contractions of the cardiac muscles.

What is an **organ system**?

An organ system is a group of organs working together to perform a vital body function. There are twelve major organ systems in the human body.

Organ Systems and Their Functions

Organ System	Components	Functions
Cardiovascular and circulatory	Heart, blood, and blood vessels	Transports blood throughout the body, supplying nutrients and carrying oxygen to the lungs and wastes to kidneys
Digestive	Mouth, esophagus, stomach, intestines, liver, and pancreas	Ingests food and breaks it down into smaller chemical units
Endocrine	Pituitary, adrenal, thyroid, and other ductless glands	Coordinates and regulates the activities of the body
Excretory	Kidneys, bladder, and urethra	Removes wastes from the bloodstream
Immune	Lymphocytes, macrophages, and antibodies	Removes foreign substances
Integumentary	Skin, hair, nails, and sweat glands	Protects the body
Lymphatic	Lymph nodes, lymphatic capillaries, lymphatic vessels, spleen, and thymus	Captures fluid and returns it to the cardiovascular system
Muscular	Skeletal muscle, cardiac muscle, and smooth muscle	Allows body movements
Nervous	Nerves, sense organs, brain, and spinal cord	Receives external stimuli, processes information, and directs activities
Reproductive	Testes, ovaries, and related organs	Carries out reproduction
Respiratory	Lungs, trachea, and other air passageways	Exchanges gases—captures oxygen (O_2) and disposes of carbon dioxide (CO_2)
Skeletal	Bones, cartilage, and ligaments	Protects the body and provides support for locomotion and movement

ANATOMICAL TERMINOLOGY

What is the **anatomical position**?

Anatomists universally defined the anatomical position as the body standing erect, facing forward, the feet are together and parallel to each other, the arms are at the side of the body with the palms facing forward. All directional terms that describe

7

the relationship of one body part to another assume the body is in the anatomical
position.

What are the commonly used **directional terms** to describe the location of one body part in relation to another body part?

Standard directional terms are used to describe the location of one body part in
relation to another body part. Most directional terms occur as pairs with one term
of the pair having the opposite meaning of the other term.

Directional Terms of the Body

Term	Definition	Example
Superior (cranial or cephalic)	Toward the head	The head is superior to the neck
Inferior (caudal)	Away from the head; toward the feet	The neck is inferior to the head
Anterior (ventral)	Toward the front	The toes are anterior to the heel
Posterior (dorsal)	Toward the back	The heel is posterior to the toes
Medial	Toward the midline of the body	The nose is medial to the eyes
Lateral	Away from the midline of the body; towards the sides	The eyes are lateral to the nose
Proximal	Toward the trunk of the body; nearer the attachment of an extremity to the trunk	The shoulder is proximal to the elbow
Distal	Away from the trunk of the body; further from the attachment of an extremity to the trunk	The wrist is distal to the shoulder
Superficial (external)	Near the surface of the body	The skin is superficial to the muscles
Deep (internal)	Farther from the surface of the body	The heart is deeper than the ribs

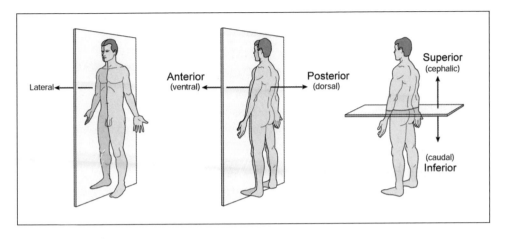

Doctors and others who study anatomy divide the human form into several directional planes to help them describe parts of the body. (From Willis, M.C. *Medical Terminology: A Programmed Learning Approach to the Language of Health Care.* Baltimore: Lippincott, Williams & Wilkins, 2002.)

What are the **two basic regions** of the **body**?

The body is divided into two basic regions: the axial and the appendicular. The axial part of the body consists of the head, neck, and trunk, including the thorax (chest), abdomen, and pelvis. The appendicular region consists of the upper and lower extremities.

What are the **divisions** of the **head and neck** regions of the body?

The head is divided into the facial region and cranium. The facial region includes the eyes, nose, and mouth. The cranium is the part of the head that covers the brain. The neck is also referred to as the cervix or cervical region.

What are the major **regions** of the **trunk**?

The trunk includes the thorax, often called chest, abdomen, and pelvis.

Major Regions of the Trunk

Region	Location
Anterior trunk	
Pectoral	The chest
Abdominal	Area between the lowest ribs and the pelvis
Pelvic	The area surrounded by the pelvic bones
Inguinal	The groin; the junction of the thighs to the anterior trunk
Posterior trunk	
Dorsum	Posterior surface of the thorax
Vertebral	Region over the vertebral column
Lumbar	Lower back region between the lowest ribs and the pelvis
Sacral	Region over the sacrum and between the buttocks
Gluteal	The buttocks

Region	Location
Lateral trunk	
Axillary	The armpits
Coxal	The hips
Inferior trunk	
Genital	External reproductive organs
Perineal	Small region between the anus and external reproductive organs

How is the **abdomen** divided into **nine regions**?

The abdomen is divided into nine regions with two vertical lines and two horizontal lines. The two vertical lines are drawn downward from the center of the collarbones. One horizontal line is placed at the lower edge of the rib cage and the other is placed at the upper edge of the hip bones. The umbilical region, containing the navel, is the center of the abdomen.

What are the areas of the **upper** and **lower extremities**?

The upper extremities and lower extremities form the appendicular region of the body. The upper extremities include the shoulders, upper arms, forearms, wrists, and hands. The lower extremities include the thighs, legs, ankles, and feet.

Major Regions of the Upper and Lower Extremities

Anatomical Term	Common Term
Upper Extremity	
Antebrachial	Forearm
Brachial	Upper arm
Antecubital	Anterior portion of the elbow joint
Cubital	Posterior portion of the elbow joint
Digital	Fingers
Palmar	Palm of the hand
Lower Extremity	
Femoral	Thigh
Patellar	Anterior portion of the knee joint
Popliteal	Posterior portion of the knee joint
Tarsal	Ankle
Pedal	Foot
Digital	Toes
Plantar	Sole of the foot

What is the **function** of the **body cavities**?

The body cavities house and protect the internal organs. There are two main body cavities: the dorsal cavity and the ventral cavity. The dorsal or posterior cavity contains the cranial cavity and the spinal cavity. The cranial cavity houses and protects the brain, while the spinal cavity houses and protects the spinal cord.

The ventral or anterior cavity is separated into the thoracic cavity and abdomino-pelvic cavity. The thoracic cavity contains the heart and lungs. It is protected by the rib cage. The abdominopelvic cavity is further divided into the abdominal cavity and the pelvic cavity. The stomach, intestines, liver, gall bladder, pancreas, spleen, and kidneys are in the abdominal cavity. The urinary bladder, internal reproductive organs, sigmoid colon, and rectum are in the pelvic cavity.

Which **structure** separates the **thoracic cavity** from the **abdominopelvic cavity**?

The diaphragm separates the thoracic cavity from the abdominopelvic cavity in the ventral cavity. It is a thin, dome-shaped sheet of muscle.

IMAGING TECHNIQUES

How do **physicians** and other health care providers **explore** the **inside of the body**?

Until the end of the 19th century, there were no noninvasive techniques to explore the internal organs of the body. Medical practitioners relied on descriptions of symptoms as the basis for their diagnoses. X-rays, discovered at the very end of the 19th century, provided the earliest technique to explore the internal organs and tissues of the body. During the 20th century, significant advances were made in the field of medical imaging to explore the internal organs.

What are **X-rays**?

X-rays are electromagnetic radiation with short wavelengths (10^{-3} nanometers) and a great amount of energy. They were discovered in 1898 by William Conrad Roentgen (1845–1923). X-rays are frequently used in medicine because they are able to pass through opaque, dense structures such as bone and form an image on a photographic plate. They are especially helpful in assessing damage to bones, identifying certain tumors, and examining the chest—heart and lungs—and abdomen.

What are **CAT** or **CT scans**?

CAT or CT scans (computer-assisted tomography or simply computerized tomography), are specialized X-rays that produce cross-sectional images of the body. An X-ray-emitting device moves around the body region being examined. At the same time, an X-ray detecting device moves in the opposite direction on the other side of the body. As these two devices move, an X-ray beam passes through the body from hundreds of different angles. Since tissues and organs absorb X-rays differently, the intensity of X-rays reaching the detector varies from position to position. A computer records the measurements made by the X-ray detector and combines them mathematically. The result is a sectional image of the body that is viewed on a screen.

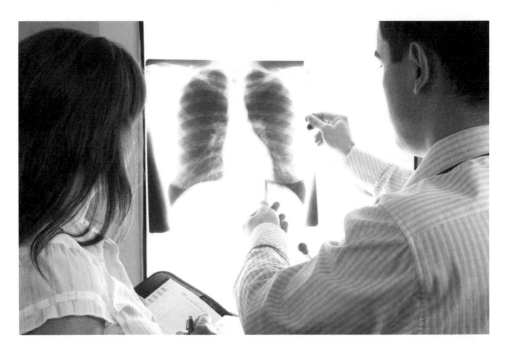

Using electromagnetic radiation known as X-rays, physicians can peer inside the human body to help them make diagnoses.
© iStockphoto.com/Christopher Pattberg Fotodesign.

How are **CT scans used** in the study of the human body?

CT scans are used to study many parts of the body, including the chest, belly and pelvis, extremities (arms and legs), and internal organs, such as pancreas, liver, gall bladder, and kidneys. CT scans of the head and brain may detect an abnormal mass or growth, stroke damage, area of bleeding, or blood vessel abnormality. Patients complaining of pain may have a CT scan to determine the source of the pain. Sometimes a CT scan will be used to further investigate an abnormality found on a regular X-ray.

Who **discovered** and pioneered the use of **CT scans**?

Dr. Allan M. Cormack (1924–1998) and Godfrey N. Hounsfield (1919–2004) independently discovered and developed computer assisted tomography in the early 1970s. They shared the 1979 Nobel Prize in Physiology or Medicine for their research. The earliest computer-assisted tomography was used to examine the skull and diseases of the brain.

What is an **advantage** of **positron emission tomography (PET imaging)** over CT scans and X-rays?

Unlike traditional X-rays and CT scans, which reveal information about the structure of internal organs, positron emission tomography (PET imaging) is an excellent technique for observing metabolic processes. Developed during the 1970s, PET imaging uses radioactive isotopes to detect biochemical activity in a specific body part.

> ## What are the disadvantages of X-rays as a diagnostic tool?
>
> A major disadvantage of X-rays as a diagnostic tool is that they provide little information about the soft tissues. Since they only show a flat, two-dimensional picture, they cannot distinguish between the various layers of an organ, some of which may be healthy while others may be diseased.

What is the **procedure** for a **PET scan**?

A patient is injected with a radioisotope, which travels through the body and is transported to the organ and tissue to be studied. As the radioisotopes are absorbed by the cells, high-energy gamma rays are produced. A computer collects and analyzes the gamma-ray emission, producing an image of the organ's activity.

How are **PET** scans used to **detect** and **treat** cancer?

PET scans of the whole body may detect cancers. While the PET scans do not provide cancer therapy, they are very useful in examining the effects of cancer therapies and treatments on a tumor. Since it is possible to observe biochemical activities of cells and tumors using PET scans, biochemical changes to tumors following treatment may be observed.

Is it possible to study **blood flow** to the **heart** or **brain**?

PET scans provide information about blood flow to the heart muscle and brain. They may help evaluate signs of coronary heart disease and reasons for decreased function in certain areas of the heart. PET scans of the brain may detect tumors or other neurological disorders, including certain behavioral health disorders. Studies of the brain using PET scans have identified parts of the brain that are affected by epilepsy and seizures, Alzheimer's disease, Parkinson's disease, and stroke. In addition, they have been used to identify specific regions of the healthy brain that are active during certain tasks.

What is **nuclear magnetic resonance (NMR)**?

Nuclear magnetic resonance (NMR) is a process in which the nuclei of certain atoms absorb energy from an external magnetic field. Scientists use NMR spectroscopy to identify unknown compounds, check for impurities, and study the shapes of molecules. This technology takes advantage of the fact that different atoms will absorb electromagnetic energy at slightly different frequencies.

What is nuclear **magnetic resonance imaging**?

Magnetic resonance imaging (MRI), sometimes called nuclear magnetic resonance imaging (NMR), is a noninvasive, nonionizing diagnostic technique. It is useful in detecting small tumors, blocked blood vessels, or damaged vertebral discs. Because it does not involve the use of radiation, it can often be used in cases where X-rays

13

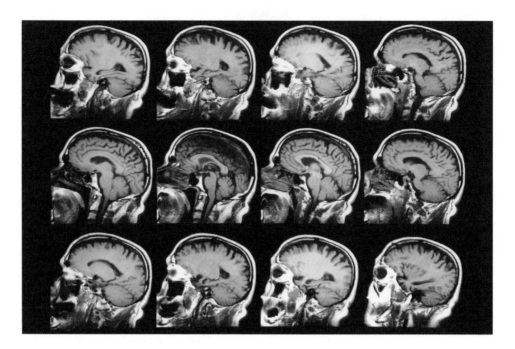

Magnetic resonance imaging (MRI) is a technique that can be less dangerous to tissues and reveal problems that X-rays miss. © iStockphoto.com/Aggressive Entertainment.

would be dangerous. Large magnets beam energy through the body, causing hydrogen atoms in the body to resonate. This produces energy in the form of tiny electrical signals. A computer detects these signals, which vary in different parts of the body and according to whether an organ is healthy or not. The variation enables a picture to be produced on a screen and interpreted by a medical specialist.

What distinguishes MRI from computerized X-ray scanners is that most X-ray studies cannot distinguish between a living body and a cadaver, while MRI "sees" the difference between life and death in great detail. More specifically, it can discriminate between healthy and diseased tissues with more sensitivity than conventional radiographic instruments like X-rays or CAT scans.

Who proposed using **magnetic resonance imaging** for **diagnostic** purposes?

The concept of using MRI to detect tumors in patients was proposed by Raymond Damadian (1936–) in a 1972 patent application. The fundamental MRI imaging concept used in all present-day MRI instruments was proposed by Paul Lauterbur (1929–) in an article published in *Nature* in 1973. Lauterbur and Peter Mansfield (1933–) were awarded the Nobel Prize in Physiology or Medicine in 2003 for their discoveries concerning magnetic resonance imaging. The main advantages of MRI are that it not only gives superior images of soft tissues (like organs), but it can also measure dynamic physiological changes in a noninvasive manner (without penetrating the body in any way). A disadvantage of MRI is that it cannot be used for every patient. For example, patients with implants, pacemakers, or cerebral aneurysm clips made of metal cannot be examined using MRI because the machine's magnet could potentially move these objects within the body, causing damage.

What is **ultrasound**?

Ultrasound, also called sonography, is another type of 3-D computerized imaging. Using brief pulses of ultrahigh frequency acoustic waves (lasting 0.01 seconds), it can produce a sonar map of the imaged object. The technique is similar to the echolocation used by bats, whales, and dolphins. By measuring the echo waves, it is possible to determine the size, shape, location, and consistency (whether it is solid, fluid-filled, or both) of an object.

Why is **ultrasound** used frequently in **obstetrics**?

Ultrasound is a very safe, noninvasive imaging technique. Unlike X-rays, sonography does not use ionizing radiation to produce an image. It gives a clear picture of soft tissues, which do not show up in X-rays. Ultrasound causes no health problems (for a mother or unborn fetus) and may be repeated as often as necessary.

Which **imaging technique** is used to **examine breast tissue** and diagnose breast diseases?

Mammography is the specific imaging technique used to examine breast tissue and diagnose breast diseases. A small dose of radiation is passed through the breast tissue. Mammography has become a very important tool in diagnosing early breast cancer. Small tumors may be visible on a mammogram years before they may be felt physically by a woman or her healthcare provider.

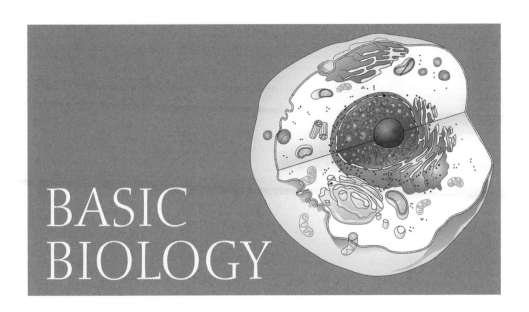

BASIC BIOLOGY

CHEMISTRY FOR BIOLOGY

Why is **chemistry important** for understanding the **human body**?

The universe and everything in it is composed of matter. Matter is anything that occupies space and has mass. The 92 naturally occurring chemical elements are the fundamental forms of matter. Twenty-six different elements are found in the human body. The continually ongoing chemical reactions in the body underlie all physiological processes of the body, including movement, digestion, the pumping of the heart, respiration, and sensory and neural processes.

What are some **important elements** in living systems?

Important elements in living systems include oxygen, carbon, hydrogen, nitrogen, calcium, phosphorus, potassium, sulfur, sodium, chlorine, magnesium, and iron. These elements are essential to life because of how they function within cells.

Common and Important Chemical Elements in the Human Body

Element	% of Humans by Weight	Functions in Life
Oxygen	65	Part of water and most organic molecules; essential to physiological processes
Carbon	18	Basic component of organic molecules
Hydrogen	10	Part of most organic molecules and water
Nitrogen	3	Component of proteins and nucleic acids
Calcium	2	Component of bones; essential for nerves and muscles; blood clotting
Phosphorus	1	Part of cell membranes and energy storage molecules; component of bones, teeth, and nervous tissue

Element	% of Humans by Weight	Functions in Life
Potassium	0.3	Important for nerve function, muscle contraction, and water-ion balance in body fluids
Sulfur	0.2	Structural component of some proteins
Sodium	0.1	Primary ion in body fluids; essential for nerve function
Chlorine	0.1	Major ion in body fluids
Magnesium	Trace	Cofactor for enzymes; important to muscle contraction and nerve transmission
Iron	Trace	Basic component of hemoglobin

Why do we **die without oxygen**?

Most living organisms are aerobic; that is, they require oxygen to complete the total breakdown of glucose for the production of adenosine triphosphate (ATP), the energy for life. Many people think that humans need oxygen to breathe, but actually people need oxygen to recycle the spent electrons and hydrogen ions (H^+) produced as byproducts of aerobic respiration.

Why is **water important** to living organisms?

Water serves many purposes in the functioning of the human body. For example, in digestion it serves as a solvent to break down large compounds into smaller ones. Water is also a transporter of nutrients, waste products, blood, and materials within cells. Water is very important in temperature regulation through perspiration and evaporation. Finally, water is a main component of synovial fluid, the lubricating fluid that helps joints move smoothly and easily.

What is the **water content** of **various tissues** of the human body?

Water accounts for approximately 62 percent of the total body weight of a human. It is found is every tissue.

Tissue	% Body Weight	% Water	Quarts of Water (Liters)
Muscle	41.7	75.6	23.35 (22.1)
Skin	18	72	9.58 (9.07)
Blood	8	83	4.91 (4.65)
Skeletal	15.9	22	2.59 (2.45)
Brain	2	74.8	1.12 (1.05)
Liver	2.3	68.3	1.16 (1.1)

Tissue	% Body Weight	% Water	Quarts of Water (Liters)
Intestines	1.8	74.5	0.99 (0.94)
Fat tissue	8.5	10	0.74 (0.7)
Lungs	0.7	79	0.41 (0.39)
Heart	0.5	79.2	0.3 (0.28)
Kidneys	0.4	82.7	0.24 (0.23)
Spleen	0.2	75.8	0.12 (0.11)

What is the **pH scale**?

The pH scale is the measurement of the hydrogen ion (H^+) concentration in an aqueous solution. It is used to measure the acidity or alkalinity of a solution. The pH scale ranges from 0 to 14. A neutral solution has a pH of 7; a solution with a pH greater than 7 is basic (or alkaline); and a solution with a pH less than 7 is acidic. The lower the pH is, the more acidic the solution is. As the pH scale is logarithmic, each whole number drop on the scale represents a tenfold increase in acidity (the concentration of H^+ increases tenfold).

Examples of pH Values

Example of Solutions	pH Value
Hydrochloric acid, battery acid	0.0
Gastric juice (digestive juice of the stomach)	1.2–3.0
Lemon juice	2.3
Grapefruit juice, vinegar, wine	3.0
Carbonated soft drink	3.0–3.5
Orange juice	3.5
Vaginal fluid	3.5–4.5
Tomato juice	4.2
Coffee	5.0
Urine	4.6–8.0
Saliva	6.35–6.85
Cow's milk	6.8
Distilled (pure) water	7.0
Blood	7.35–7.45
Semen (fluid containing sperm)	7.2–7.6
Cerebrospinal fluid (fluid associated with nervous system)	7.4
Pancreatic juice (digestive juice of the pancreas)	7.1–8.2
Bile (liver secretion that aids fat digestion)	7.6–8.6
Milk of magnesia	10.5
Lye	14.0

BIOLOGICAL COMPOUNDS

What are the **major bioorganic molecules** in humans?

The major bioorganic molecules are carbohydrates, lipids, proteins, and nucleic acids. These molecules are characteristic of life and have basic roles such as storing and producing energy, providing structural materials, or storing hereditary information.

What are **carbohydrates**?

Carbohydrates are organic compounds composed of carbon, hydrogen, and oxygen. The general chemical formula for carbohydrates is CH_2O, indicating there is twice as much hydrogen as oxygen. Carbohydrates are the major source of energy for cells and cellular activities.

How are **carbohydrates classified**?

Carbohydrates are classified in several ways. Monosaccharides (single unit sugars) are grouped by the number of carbon molecules they contain: triose has three, pentose has five, and hexose has six. Carbohydrates are also classified by their overall length (monosaccharide, disaccharide, polysaccharide) or function. Examples of functional definitions are storage polysaccharides (glycogen and starch), which store energy, and structural polysaccharides (cellulose and chitin).

What are some of the **uses** of **carbohydrates** by the **body**?

Carbohydrates are mainly used as an energy source by the body. Different carbohydrates have different functions. The following chart identifies some common carbohydrates and their uses.

Carbohydrate Name	Type	Use by the Body
Deoxyribose	Monosaccharide	DNA; constituent of hereditary material
Fructose	Monosaccharide	Important in cellular metabolism of carbohydrates
Galactose	Monosaccharide	Found in brain and nerve tissue
Glucose	Monosaccharide	Main energy source for the body
Ribose	Monosaccharide	Constituent of RNA
Lactose	Disaccharide	Milk sugar; aids the absorption of calcium
Sucrose	Disaccharide	Produces glucose and fructose upon hydrolysis
Cellulose	Polysaccharide	Not digestible by the body, but is an important fiber that provides bulk for the proper movement of food through the intestines
Glycogen	Polysaccharide	Stored in the liver and muscles until needed as energy source and is then converted to glucose
Heparin	Polysaccharide	Prevents excessive blood clotting
Starch	Polysaccharide	Chief food carbohydrate in human nutrition

What are **lipids**?

Lipids are organic compounds composed mainly of carbon, hydrogen, and oxygen, but they also may contain other elements, such as phosphorus and nitrogen. Lipids usually have more than twice as many hydrogen atoms as oxygen atoms. They are insoluble in water, but can be dissolved in certain organic solvents such as ether, alcohol, and chloroform. Lipids include fats, oils, phospholipids, steroids, and prostaglandins.

What is the **difference** between **fats** and **lipids**?

Fats are one category of lipids. Each fat molecule is comprised of a glycerol (alcohol) molecule and at least one fatty acid (a hydrocarbon chain with an acid group attached). Fats are energy-rich molecules important as a source of reserve food for the body. They are

People often associate the word "carbohydrates" with a fatty diet that makes one overweight, but sensible intake of carbs is an essential source of energy. © iStockphoto.com/Maica.

stored in the body in the form of triacylglycerols, also known as triglycerides. Fats also provide the body with insulation, protection, and cushioning.

What is **cholesterol**?

Cholesterol belongs to a category of lipids known as steroids. Steroids have a unique chemical structure. They are built from four carbon-laden ring structures that are fused together. The human body uses cholesterol to maintain the strength and flexibility of cell membranes. Cholesterol is also the molecule from which steroid hormones and bile acids are built.

What is an **enzyme**?

An enzyme is a protein that acts as a biological catalyst. It decreases the amount of energy needed (activation energy) to start a metabolic reaction. Different enzymes work in different environments due to changes in temperature and acidity. For example, the amylase that is active in the mouth cannot function in the acidic environment of the stomach; pepsin, which breaks down proteins in the stomach, cannot function in the mouth. Without enzymes, the stomach would not be able to harvest energy and nutrients from food.

What are some of the most **common enzyme deficiencies**?

Lactose intolerance, a condition that results from the inability to digest lactose—the sugar present in milk—is one of the most common enzyme deficiencies. Glucose-6-phosphate dehydrogenase deficiency is a more serious enzyme deficiency that is linked to the bursting of red blood cells (hemolysis). This deficiency is found in more than 200 million people, mainly Mediterranean, West African, Middle Eastern, and Southeast Asian populations.

What are **proteins** and what is their **purpose**?

Proteins are large, complex molecules composed of smaller structural subunits called amino acids. All proteins contain carbon, hydrogen, oxygen, and nitrogen, and sometimes sulfur, phosphorus, and iron. Human life could not exist without proteins. The enzymes that are required for all metabolic reactions are proteins. Proteins also are important to structures like muscles, and they act as both transporters and signal receptors.

Type of Protein	Examples of Functions
Defensive	Antibodies that respond to invasion
Enzymatic	Increase the rate of reactions; build and breakdown molecules
Hormonal	Insulin and glucagon, which control blood sugar
Receptor	Cell surface molecules that cause cells to respond to signals
Storage	Store amino acids for use in metabolic processes
Structural	Major components of muscles, skin, hair
Transport	Hemoglobin carries oxygen from lungs to cells

CELLS

What is the **chemical composition** of a typical **mammalian cell**?

Molecular Component	% of Total Cell Weight
Water	70
Proteins	18
Phospholipids and other lipids	5
Miscellaneous small metabolites	3
Polysaccharides	2
Inorganic ions (sodium, potassium, magnesium, calcium, chlorine, etc.)	1
RNA	1.1
DNA	0.25

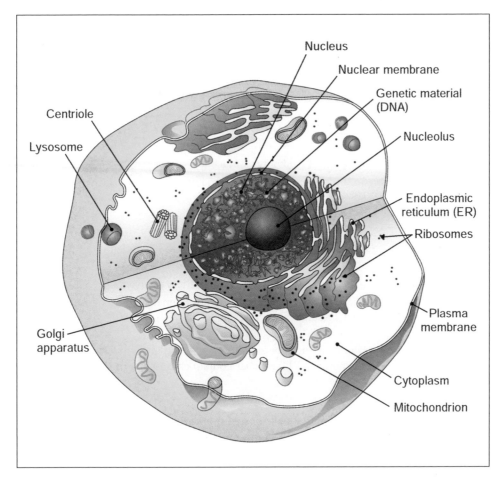

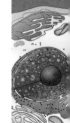

Organelles are structures within a cell that serve specific purposes. (Cohen, B. J., and Wood, D.L. *Memmler's The Human Body in Health and Disease*. 9th Ed. Philadelphia: Lippincott, Williams & Wilkins, 2000.)

What are **organelles**?

Organelles—frequently called "little organs"—are found in all eukaryotic cells; they are specialized, membrane-bound, cellular structures that perform a specific function. Eukaryotic cells contain several kinds of organelles, including the nucleus, mitochondria, chloroplasts, endoplasmic reticulum, and Golgi apparatus.

What are the **major components** of the **eukaryotic cell**?

Structure	Description
Cell Nucleus	
Nucleus	Large structure surrounded by double membrane
Nucleolus	Special body within nucleus; consists of RNA and protein
Chromosomes	Composed of a complex of DNA and protein known as chromatin; resemble rodlike structures after cell division

23

Structure	Description
Cytoplasmic Organelles	
Plasma membrane	Membrane boundary of living cell
Endoplasmic reticulum (ER)	Network of internal membranes extending through cytoplasm
Smooth endoplasmic reticulum	Lacks ribosomes on the outer surface
Rough endoplasmi reticulum	Ribosomes stud outer surface
Ribosomes	Granules composed of RNA and protein; some attached to ER and some are free in cytosol
Golgi complex	Stacks of flattened membrane sacs
Lysosomes	Membranous sacs (in animals)
Vacuoles	Membranous sacs (mostly in plants, fungi, and algae)
Microbodies (e.g., peroxisomes)	Membranous sacs containing a variety of enzymes
Mitochondria	Sacs consisting of two membranes; inner membrane is folded to form cristae and encloses matrix
Plastids (e.g., chloroplasts)	Double membrane structure enclosing internal thylakoid membranes; chloroplasts contain chlorophyll in thylakoid membranes
The Cytoskeleton	
Microtubules	Hollow tubes made of subunits of tubulin protein
Microfilaments	Solid, rod-like structures consisting of actin protein
Centrioles	Pair of hollow cylinders located near center of cell; each centriole consists of nine microtubule triplets (9×3 structure)
Cilia	Relatively short projections extending from surface of cell; covered by plasma membrane; made of two central and nine peripheral microtubules ($9 + 2$ structure)
Flagella	Long projections made of two central and nine peripheral microtubules ($9 + 2$ structure); extend from surface of cell; covered by plasma membrane

Do all **human cells** have a **nucleus**?

Most eukaryotic cells have a single organized nucleus. The red blood cell is the only mammalian cell that does not have a nucleus.

What are the **main components** of the **nucleus**?

The nucleus, the largest organelle in an eukaryotic cell, is the repository for the cell's genetic information and the control center for the expression of that information. The boundary around the nucleus consists of two membranes (an inner one and an outer one) that form the nuclear envelope. Nuclear pores are small openings in the nuclear envelope that permit molecules to move between the nucleus and the cytoplasm. The nucleolus is a prominent structure within the nucleus. The nucleoplasm is the viscous liquid contained within the nucleus. In addition, the DNA-bearing chromosomes of the cell are found in the nucleus.

How much **DNA** is in a typical **human cell**?

If the DNA (deoxyribonucleic acid) molecules in a single human cell were stretched out and laid end-to-end they would measure approximately 6.5 feet (2 meters). The average human body contains 10 to 20 billion miles (16 to 32 billion kilometers) of DNA distributed among trillions of cells. If the total DNA in all the cells from one human were unraveled, it would stretch to the sun and back more than 500 times.

How is **DNA organized** in the **nucleus**?

Within the nucleus, DNA (deoxyribonucleic acid) is organized with proteins into a fibrous material called chromatin. As a cell pre-

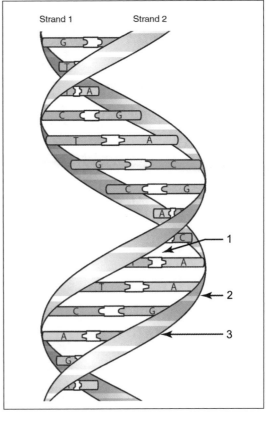

A DNA molecule is comprised of sequences of four molecules. In the above chart, C = Cytosine; G = Guanine; A = Adenine; T = Thymine; 1 = Hydrogen bond; 2 = Phosphate group; 3 = Deoxyribose sugar. (From Premkumar K. *The Massage Connection Anatomy and Physiology.* Baltimore: Lippincott, Williams & Wilkins, 2004.)

pares to divide or reproduce, the thin chromatin fibers condense, becoming thick enough to be seen as separate structures, which are called chromosomes.

What is a **chromosome**?

A chromosome is the threadlike part of a cell that contains DNA and carries the genetic material of a cell. In prokaryotic cells chromosomes consist entirely of DNA and are not enclosed in a nuclear membrane. In eukaryotic cells the chromosomes are found within the nucleus and contain both DNA and RNA (ribonucleic acid).

What are **lysosomes**?

Lysosomes, first observed by Belgian biochemist Christian de Duve (1917–) in the early 1950s, are single, membrane-bound sacs that contain digestive enzymes. The digestive enzymes break down all the major classes of macromolecules including proteins, carbohydrates, fats, and nucleic acids. Throughout a cell's lifetime, the lysosomal enzymes digest old organelles to make room for newly formed organelles.

Multiple DNA molecules are entwined to form a chromosome; eukaryotic cells also contain RNA. *Anatomical Chart Co.*

The lysosomes allow cells to continually renew themselves and prevent the accumulation of cellular toxins.

What are **mitochondria**?

A mitochondrion (singular form) is a self-replicating, double-membraned body found in the cytoplasm of all eukaryotic cells. The outer membrane of a mitochondrion is smooth, while the inner membrane is folded into numerous layers that are called cristae. Mitochondria are the location for much of the metabolism necessary for protein synthesis, and for the production of both ATP and lipids.

How **many mitochondria** are there in a **cell**?

The number of mitochondria varies according to the type of cell. The number ranges between 1 and 10,000, but averages about 200. Each cell in the human liver has over 1,000 mitochondria. Cells with high energy requirements, such as muscle cells, may have many more mitochondria.

What is **ATP**?

ATP (adenosine triphosphate) is the universal energy currency of a cell. Its secret lies in its structure. ATP contains three negatively charged phosphate groups. When the bond between the outermost two phosphate groups is broken, ATP becomes ADP (adenosine diphosphate). This reaction releases 7.3 kilocalories/mole of ATP, which is a great deal of energy by cell standards.

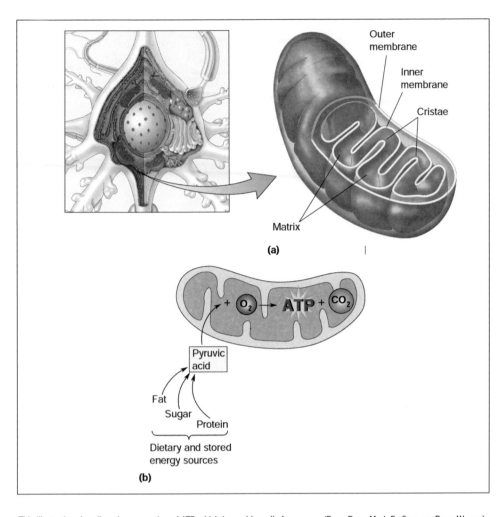

This illustration describes the generation of ATP, which is used by cells for energy. (From Bear, Mark F., Connors, Barry W., and Paradiso, Michael A. *Neuroscience: Exploring the Brain*. 2nd Ed. Philadelphia: Lippincott, Williams & Wilkins, 2001.)

How much ATP does the human body use?

Each cell in the human body is estimated to use between 1 and 2 billion ATPs per minute, which comes to roughly 1×10^{23} for a typical human body. In the span of 24 hours, the body's cells produce about 441 pounds (200 kilograms) of ATP.

How many cells are in the human body?

Scientists estimate there are between 50 and 100 trillion cells in the human body.

What is the average lifespan of various cells in the human body?

The human body is self-repairing and self-replenishing. According to one estimate, almost 200 billion cells die each hour. In a healthy body, dying cells are simultaneously replaced by new cells.

What is the function of the Golgi apparatus?

The Golgi apparatus (frequently called the Golgi body), first described in 1898 by Italian histologist Camillo Golgi (1843–1926), is a collection of flattened stacks of membranes. It serves as the packaging center for cell products. It collects materials at one place in the cell, and packages them into vesicles for use elsewhere in the cell or for transportation out of the cell.

Cell Type	Average Lifespan
Blood cells: Red blood cells	120 days
Blood cells: Lymphocytes	Over 1 year
Blood cells: Other white cells	10 hours
Blood cells: Platelets	10 days
Bone cells	25–30 years
Brain cells*	Lifetime
Colon cells	3–4 days
Liver cells	500 days
Skin cells	19–34 days
Spermatozoa	2–3 days
Stomach cells	2 days

*Brain cells are the only cells that do not divide further during a person's lifetime. They either last the entire lifetime, or if they die during a person's lifetime they are not replaced.

TISSUES

Where is epithelial tissue found?

Epithelial tissue, also called epithelium (from the Greek *epi,* meaning "on," and *thele,* meaning "nipple"), covers every surface, both external and internal, of the body. The outer layer of the skin, the epidermis, is one example of epithelial tissue. Other examples of epithelial tissue are the lining of the lungs, kidney tubules, and the inner surfaces of the digestive system, including the esophagus, stomach, and intestines. Epithelial tissue also includes the lining of parts of the respiratory system.

What are the different shapes and functions of the epithelium?

Epithelial tissue consists of densely packed cells. It is either simple or stratified, based on the number of cell layers. Simple epithelium has one layer of cells, while stratified epithelium has multiple layers. Epithelial tissue may have squamous-, cuboidal-, or columnar-shaped cells. Squamous cells are flat, square cells. Cuboidal

cells form a box or cube. Columnar cells are stacked, forming a column taller than they are wide. There are two surfaces to epithelial tissue: one side is firmly attached to the underlying structure, while the other forms the lining. The epithelium forms a barrier, allowing the passage of certain substances, while impeding the passage of other substances.

Where are **different types** of **epithelial tissues** found in the body?

The different types of epithelial tissue are located in different parts of the body according to their specialization.

Type of Epithelial Tissue	Major Locations	Major Functions
Simple squamous	Lining of lymph vessels, blood vessels, heart, glomerular capsule in kidneys, alveoli (air sacs in lungs), serous membranes lining peritoneal, pleural, pericardial, and scrotal cavities	Permits diffusion or filtration through selectively permeable surfaces
Simple cuboidal	Lining of many glands and their ducts, surface of ovaries, inner surface of eye lens, pigmented epithelium of eye retina	Secretion and absorption
Simple columnar	Stomach, intestines, digestive gland, and gall bladder	Secretion, absorption, protection, lubrication; cilia and mucus combine to sweep away foreign substances
Stratified squamous	Epidermis, vagina, mouth and esophagus, anal canal, distal end of urethra	Protection
Stratified cuboidal	Ducts of sweat glands, sebaceous glands, and developing epithelium in ovaries and testes	Secretion
Stratified columnar	Moist surfaces such as larynx, nasal surface of soft palate, parts of pharynx, urethra, and excretory ducts of salivary and mammary glands	Secretion and movement

What is the **basement membrane**?

The basement membrane is a thin layer composed of tiny fibers and nonliving polysaccharide material produced by epithelial cells. It anchors the epithelial tissue to the underlying connective tissue. The basement membrane provides elastic support and acts as a partial barrier for diffusion and filtration.

29

Do **epithelial tissues** contain **blood vessels**?

Epithelial tissues are avascular, meaning they do not contain blood vessels. Oxygen and other nutrients diffuse through the permeable basement membranes from capillaries in the underlying connective tissue, while wastes diffuse into connective tissue capillaries.

How often is the **epithelium replaced**?

Epithelial cells are constantly being replaced and regenerated during an individual's lifetime. The epidermis (outer layer of the skin) is renewed every two weeks, while the epithelial lining of the stomach is replaced every two to three days. The lining of the respiratory tract is only replaced every five to six weeks. The liver, a gland consisting of epithelial tissue, easily regenerates after portions are removed surgically.

Which kinds of **epithelial tissues cannot be classified** easily as typical epithelia?

Pseudostratified columnar epithelium, transitional epithelium, and glandular epithelium cannot be classified as easily as typical epithelium. Pseudostratified columnar epithelium, found in the trachea, bronchi and large bronchioles, and parts of the male reproductive tract, is characterized by the fact that all of its cells are in contact with the basement membrane, but not all of the cells reach the surface. It is called pseudostratified because it gives the false ("pseudo") impression that it is a multilayered stratification, since the nuclei of the cells appear to be at several different levels.

Transitional epithelium lines the urinary tract, including the ureters, urinary bladder, urethra, and calyxes of the kidneys. The cells vary in shape depending on the amount of fluid the organ contains. For example, when the urinary bladder contains a large quantity of urine, the cells are stretched out and assume a flat, squamous appearance. When the bladder is empty, the cells have a cuboidal or slightly columnar shape.

Glandular epithelium cells are specialized for the synthesis, storage, and secretion of chemical substances, such as saliva or digestive juices. These glands are called exocrine glands.

What is a **gland**?

Glands are secretory cells or multicellular structures that are derived from epithelium and often stay connected to it. They are specialized for the synthesis, storage, and secretion of chemical substances. Glands are classified as either endocrine or exocrine glands. Endocrine glands do not have ducts, but release their secretions directly into the extracellular fluid. The secretions pass into capillaries and are then transported by the bloodstream to target cells elsewhere in the body.

Exocrine glands have ducts that carry the secretions to some body surface. Mucus, saliva, perspiration, earwax, oil, milk, and digestive enzymes are examples of exocrine secretions.

What is the only unicellular exocrine gland in the human body?

The goblet, or mucus cell, is the only unicellular exocrine gland in the human body. It is found in the lining of the intestines and other parts of the digestive system, the respiratory tracts, and the conjunctiva of the eye. Goblet cells produce a carbohydrate-rich glycoprotein called mucin, which is then secreted in the form of mucus, a thick, lubricating fluid.

How are **exocrine glands** classified?

Exocrine glands may be unicellular or multicellular. Multicellular exocrine glands may be either simple or compound glands. Simple glands are glands with only one unbranched duct, while those with more than one branch are compound glands.

What is the **unique characteristic** of **connective tissue**?

The cells of connective tissue are spaced widely apart and are scattered through a nonliving extracellular material called a matrix. The matrix, which varies in different types of connective tissue, may be a liquid, jelly, or solid.

What are the **major types** of **connective tissue** and their **function**?

The major types of connective tissue are: 1) loose connective tissue; 2) adipose tissue; 3) blood; 4) collagen, sometimes called fibrous or dense connective tissue; 5) cartilage; and 6) bone.

Loose connective tissue, also called areolar tissue (from the Latin *areola,* meaning "open place"), is a mass of widely scattered cells whose matrix is a loose weave of fibers. Many of the fibers are strong protein fibers called collagen. Loose connective tissue is found beneath the skin and between organs. It is a binding and packing material whose main purpose is to provide support to hold other tissues and organs in place.

Adipose tissue consists of adipose cells in loose connective tissue. Each adipose cell stores a large droplet of fat that swells when fat is stored and shrinks when fat is used to provide energy. Adipose tissue provides padding, absorbs shocks, and insulates the body to slow heat loss.

Blood is a loose connective tissue whose matrix is a liquid called plasma. Blood consists of red blood cells (erythrocytes), white blood cells (leukocytes), and platelets (thrombocytes), which are tiny pieces of bone marrow cell. Plasma also contains water, salts, sugars, lipids, and amino acids. Blood is approximately 55 percent plasma and 45 percent formed elements. Blood transports substances from one part of the body to another and plays an important role in the immune system.

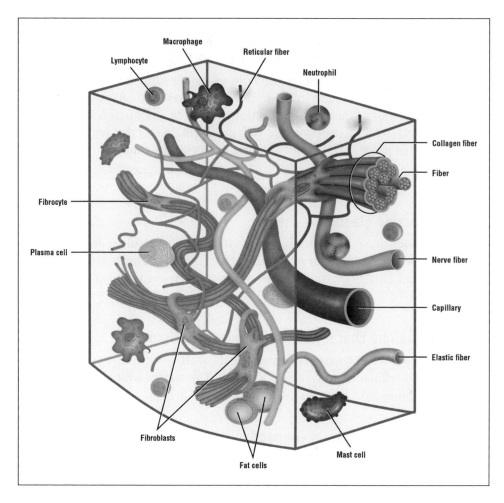

The human body has many types of connective tissues in the form of fibers and individual cells. (Image from Eroschenko, V.P., Ph.D. *di Fiore's Atlas of Histology. With Functional Correlations.* 9th Ed. Baltimore: Lippincott, Williams & Wilkins, 2000.)

Collagen (from the Greek *kola,* meaning "glue," and *genos,* meaning "descent") is a dense connective tissue, also known as fibrous connective tissue. It has a matrix of densely packed collagen fibers. There are two types of collagen: regular and irregular. The collagen fibers of regular dense connective tissue are lined up in parallel. Tendons, which bind muscle to bone, and ligaments, which join bones together, are examples of dense regular connective tissue. The strong covering of various organs, such as kidneys and muscle, is dense irregular connective tissue.

Cartilage (from the Latin, meaning "gristle") is a connective tissue with an abundant number of collagen fibers in a rubbery matrix. It is both strong and flexible. Cartilage provides support and cushioning. It is found between the discs of the vertebrae in the spine, surrounding the ends of joints such as knees, and in the nose and ears.

Bone is a rigid connective tissue that has a matrix of collagen fibers embedded in calcium salts. It is the hardest tissue in the body, although it is not brittle. Most

> ## Does liposuction reduce the amount of adipose tissue?
>
> L iposuction is a surgical procedure that removes adipose tissue. It is a cos- metic surgical procedure that is useful for shaping the body. However, it is not a solution for obesity, since new adipose tissue will develop.

of the skeletal system is comprised of bone, which provides support for muscle attachment and protects the internal organs.

Where is **adipose tissue found**?

Adipose tissue is abundant in the body and constitutes 18 percent of an average person's body weight. Adipose tissue is found under the skin of the groin, sides, buttocks, and breasts. It is found behind the eyeballs, surrounding the kidneys, and in the abdomen and hips.

How does **brown fat** differ from **white fat**?

White fat (or adipose tissue) stores nutrients. Brown fat, also called brown adipose tissue, consumes its nutrient stores to generate heat to warm the body. It is called brown fat because it has a deep, rich, dark color that is derived from the numerous mitochondria in each individual cell. Brown adipose tissue is found in infants and very young children between the shoulder blades, around the neck, and in the anterior abdominal wall. Older children and adults rely on shivering to warm the body.

Which types of **cancers** develop and grow in which types of **tissues**?

Different types of cancers develop and grow in the different types of tissue. Carcinomas, perhaps the most common type of cancer, are cancers of the epithelial tissue. Sarcomas are cancers arising in the muscle and connective tissue. Leukemias are cancers of the blood. Lymphomas are cancers of the reticular connective tissue.

Is all the **cartilage** in the body **the same**?

There are three types of cartilage in the human body: 1) hyaline cartilage; 2) elastic cartilage; and 3) fibrocartilage. Hyaline cartilage (from the Greek *hyalos,* meaning "glass") is the most common type of cartilage in the body. It has a translucent, pearly, blue-white appearance resembling glass. Hyaline cartilage provides stiff but flexible support and reduces friction between bony surfaces. It is found between the tips of the ribs and the bones of the sternum, at the end of the long bones, at the tip of the nose, and throughout the respiratory passages.

Elastic cartilage is similar to hyaline cartilage except it is very flexible and resilient. It is ideal for areas that need repeated bending and stretching. Elastic cartilage forms the external flap of the outer ear and is found in the auditory canal and epiglottis.

Why are cartilage transplants successful?

Cartilage does not contain blood vessels. Oxygen, nutrients, and cellular wastes diffuse through the selectively permeable matrix. Cartilage transplants are successful because foreign proteins in the transplanted cells do not have a way to enter the host body's circulation and cause an immune response. However, since there are no blood vessels in cartilage, the healing process is slower than for other tissues.

Fibrocartilage is often found where hyaline cartilage meets a ligament or tendon. It is found in the pads of the knees, between the pubic bones of the pelvis, and between the spinal vertebrae. It prevents bone-to-bone contact.

What **condition** is caused by the **accumulation of fluid** in **loose connective tissue**?

Edema is the accumulation of fluid in loose connective tissue. It is characterized by swelling of the affected area.

What are the **three types** of **muscle tissue**?

There are three types of muscle tissue in the body: 1) smooth muscle; 2) skeletal muscle; and 3) cardiac muscle. Muscle tissue, consisting of bundles of long cells called muscle fibers, is specialized for contraction. It enables body movements, as well as the movement of substances within the body.

Does **exercise increase** the number of **muscle cells**?

Adults have a fixed number of skeletal muscle cells, so exercise does not increase their number. Exercise, however, does enlarge the existing skeletal muscle cells.

What type of **cell** is found in **nerve tissue**?

Neurons are specialized cells that produce and conduct "impulses," or nerve signals. Neurons consist of a cell body, which contains a nucleus and two types of cytoplasmic extensions, dendrites, and axons. Dendrites are thin, highly branched extensions that receive signals. Axons are tubular extensions that transmit nerve impulses away from the cell body, often to another neuron. Nerve tissue also has supporting cells, called neuroglia or glial cells, which nourish the neurons, insulate the dendrites and axons, and promote quicker transmission of signals.

How many **different types** of **neurons** are found in nerve tissue?

There are three main types of neurons: 1) sensory neurons; 2) motor neurons; and 3) interneurons (also called association neurons). Sensory neurons conduct impuls-

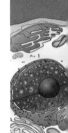

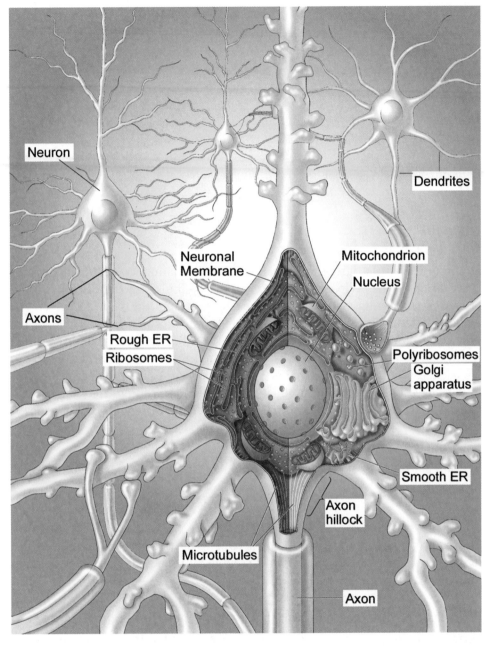

Neuron

Dendrites

Neuronal Membrane

Mitochondrion

Nucleus

Axons

Rough ER

Ribosomes

Polyribosomes

Golgi apparatus

Smooth ER

Axon hillock

Microtubules

Axon

The structure of a basic nerve cell. (From Bear, Mark F., Connors, Barry W., and Paradiso, Michael A. *Neuroscience: Exploring the Brain*. 2nd Ed. Philadelphia: Lippincott, Williams & Wilkins, 2001.)

es from sensory organs (eyes, ears, and the surface of the skin) into the central nervous system. Motor neurons conduct impulses from the central nervous system to muscles or glands. Interneurons are neither sensory neurons nor motor neurons. They permit elaborate processing of information to generate complex behaviors. Interneurons comprise the majority of neurons in the central nervous system.

Which types of tissue have the greatest capacity to regenerate?

Epithelial and connective tissues have the greatest capacity to regenerate. In small wounds and injuries, the epithelial and connective tissues often heal with normal tissue. The ability of muscle tissue to regenerate is very limited. Fibrous connective tissue often replaces damaged muscle tissue. As a consequence, the organ involved loses all or part of its ability to function. Nerve tissue has even less capacity to regenerate. Although neurons outside the brain and spinal cord sometimes regenerate at a very slow pace, most brain and spinal cord injuries result in permanent damage.

What is **myelin**?

Myelin is a white, fatty substance that forms an insulating wrapping around large nerve axons. In the peripheral nervous system, myelin is formed by Schwann cells (a type of supporting cell) that wrap repeatedly around the axon. In the central nervous system, myelin is formed by repeated wrappings of processes of oligodendrocytes (a different type of supporting cell). The process of each cell forms part of the myelin sheath. The space between the myelin from individual Schwann cells or oligodendrocyte processes is a bare region of the axon called the node of Ranvier. Nerve conduction is faster in myelinated fibers because it jumps from one node of Ranvier to the next. For this reason, it is called saltatory (jumping) conduction.

What are the **longest cells** in the body?

Neurons are the longest cells in the body. Some neurons are 39 inches (99 centimeters) long.

What is the **largest nerve** in the human body?

The sciatic nerve, running from the spinal cord to the back of each leg, is the largest in the body. It is approximately 0.78 inches (1.98 centimeters) in diameter, or about as thick as a lead pencil.

Which type of **tissue** accounts for the greatest amount of **body weight**?

Muscle tissue accounts for approximately 50 percent of body weight and connective tissue accounts for 45 percent of total body weight. The remaining 5 percent is divided between epithelium and glands (3 percent) and neural tissue (2 percent). Tissues combine to form all the organs and systems of the human body.

Is it possible to **repair damaged tissue**?

Tissue responds to injury or other damage with a two-step process: 1) inflammation; and 2) regeneration to restore homeostasis. Inflammation, or the inflammatory response, produces swelling, redness, warmth, and pain in the area of injury. The

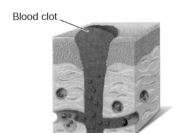

Blood clot

Immediately: Blood clot and debris fill the cut

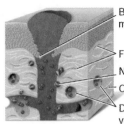

Basal epithelial cells migrate around wound

Fibroblasts

Neutrophil

Collagen fibers

Dilated blood vessels

2-3 hours: Early inflammation closes the edges

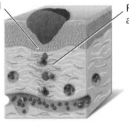

Epithelial growth

Fibroblastic activity

2-3 days: Macrophages remove blood clot. Increased fibroblastic activity and epithelial growth close gap

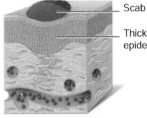

Scab

Thickening of epidermis

10-14 days: Scab formation: epithelial covering is complete and edges of wound unite by fibrous tissue; however, the wound is still weak

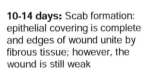

Weeks: The scar tissue is still hyperemic; union of edges is good but not full strength

Months-Years: Very little or no scars; collagen tissue remodelled by enzymes; normal blood flow

The process of damaged epithelial tissue scarring and healing. (From Premkumar K. *The Massage Connection Anatomy and Physiology*. Baltimore: Lippincott, Williams & Wilkins, 2004.)

injured area is isolated while damaged cells and dangerous microorganisms are destroyed. During the second process, regeneration, the damaged tissues are replaced or repaired to restore normal function. Regeneration begins while the cleanup processes of inflammation are still in process.

What is **pus**?

Lysosomes are responsible for releasing enzymes that destroy the injured cells and attack surrounding tissues. Pus is an accumulation of debris, fluid, dead and dying cells, and necrotic tissue. An abscess is an accumulation of pus in an enclosed tissue space.

How does a **scar** form?

A scar forms when the dense mass of fibrous connective tissue that fills in the gap after an injury is deep or large. A scar also may form when the cell damage was extensive and the dense fibrous mass remains and is not replaced by normal tissue.

MEMBRANES

What are the **four types** of **membranes**?

Membranes are thin layers of epithelial tissue usually bound to an underlying layer of connective tissue. Membranes cover, protect, or separate other structures or tissues in the body. The four types of membranes are: 1) cutaneous membranes; 2) serous membranes; 3) mucous membranes; and 4) synovial membranes.

The cutaneous membrane is skin. Skin consists of a layer of stratified squamous epithelium (epidermis) firmly attached to a thick layer of dense connective tissue (dermis). It differs from other membranes because it is exposed to air and is dry.

The serous membranes (or serosae) consist of simple squamous epithelium (a mesothelium) supported by a layer of connective tissue (areolar). These moist membranes line the closed, internal divisions of the ventral body cavity. The three types of serous membranes are: 1) the pleura, lining the pleural cavities and covering the lungs; 2) the peritoneum, lining the peritoneal cavity and covering the abdominal organs; and 3) the pericardium, lining the pericardial cavity and covering the heart.

The mucous membranes (or mucosae) consist of epithelial tissue (usually stratified squamous or simple columnar epithelia) on a layer of loose connective tissue called the lamina propria (from the Latin, meaning "one's own layer"). The mucosae line the body cavities that open to the exterior, such as the digestive, respiratory, reproductive, and urinary tracts. These membranes are kept moist by bodily secretions.

Synovial membranes are composed of connective tissue. They surround the cavity of joints, filling the space with the synovial fluid that they make. The synovial fluid lubricates the ends of the bones allowing them to move freely.

What **conditions** result from the **build-up of fluids** in the ventral body cavity?

An abnormal build-up of fluids in the ventral body cavity may be caused by infection or chronic irritation. Each type of serous membrane may be affected by inflammation and infection. Pleurisy is an inflammation of the pleural cavity; pericarditis is an inflammation of the pericardium; and peritonitis is an inflammation of the peritoneum.

HOMEOSTASIS

What is **homeostasis**?

Homeostasis (from the Greek *homois,* meaning "same," and *stasis,* meaning "standing still") is the state of inner balance and stability maintained by the human body despite constant changes in the external environment. Nearly everything that occurs in the body helps to maintain homeostasis, from kidneys filtering the blood and removing a carefully regulated amount of water and wastes to the lungs working together with the heart, blood vessels, and blood to distribute oxygen throughout the body and remove wastes.

Who coined the term "**homeostasis**"?

Walter Bradford Cannon (1871–1945), who elaborated on Claude Bernard's (1813–1878) concept of the *milieu intérieur* (interior environment), used the term "homeostasis" to describe the body's ability to maintain a relative constancy in its internal environment.

What are the **three components** necessary to **maintain homeostasis**?

The three components of homeostasis are sensory receptors, integrators, and effectors. These three components interact to maintain the state of homeostasis. Sensory receptors are cells that can detect a stimulus that signals a change in the environment. The brain is the integrator that processes the information and selects a response. Muscles and glands are effectors that carry out the response.

How does the term "**negative feedback**" apply to homeostasis?

Negative feedback is a cellular process that works in a way that is similar to the manner in which an air conditioner operates: an air conditioner is set to a specific temperature, and the air conditioner shuts off when the surrounding air reaches the set temperature. Negative feedback is part of the homeostatic process through which cells conserve energy by synthesizing products only for their immediate needs.

What is an example of a "**negative feedback**" to **maintain homeostasis**?

Maintaining the normal blood sugar level of the body is an example of negative feedback. When the blood sugar level decreases, the body responds to raise the level. If the blood sugar level increases, the body acts to lower the level. Each of these responses is a negative action, since the response does the opposite of the initial stimulus.

What is a "**positive feedback**" system?

Positive feedback systems are stimulatory, since the initial stimulus is reinforced rather than reversed. The stimulus continues to increase rapidly until the process is stopped.

Is **positive feedback commonly found** in the human body systems?

Positive feedback is relatively uncommon in the human body, since it disrupts homeostasis. For example, if there were a positive feedback response to blood sugar level decreasing, the blood sugar level would continue to decrease, without ceasing, until the person died.

What is an example of a **positive feedback control loop** in humans?

During childbirth there is an increase in the number of uterine contractions. The positive feedback response is to increase the frequency of uterine contractions. The birth of a baby stops the positive feedback response.

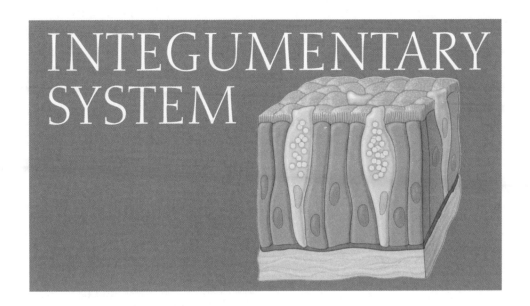

INTEGUMENTARY SYSTEM

INTRODUCTION

What **organs** are included in the **integumentary system**?

The integumentary system (from the Latin *integere,* to cover) includes skin, hair, glands, and nails. The main function of the integumentary system is to provide the body with a protective barrier between the organs inside the body and the changing environment outside.

How much **skin** does the **average person** have?

The average person is covered with about 20 square feet (1.9 square meters) of skin which weighs about 5.6 pounds (2.7 kilograms). The skin is the largest and heaviest organ in the body, representing four percent of the average weight of the human body.

How much **skin** does a person **shed** in one year?

An average man or woman sheds about 600,000 particles of skin per hour, which is approximately 1.5 pounds (680 grams) per year. Using this figure, by the age of 70, a person will have lost 105 pounds (47.6 kilograms) of skin which is equivalent to two-thirds of their entire body weight.

What **structures** are present in an average **square inch** (6.4 square centimeters) of **skin**?

The average square inch (6.4 square centimeters) of skin holds 20 feet (6.1 meters) of blood vessels, 77 feet (23.5 meters) of nerves, and more than a thousand nerve endings. In addition to blood vessels and nerves, there are 645 sweat glands, 65 hair follicles, and 97 sebaceous glands per square inch.

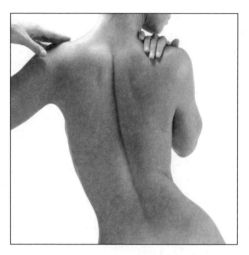

Skin is the largest human organ, with most people having about 20 square feet covering their bodies. © iStockphoto.com/Gilas

How many **bacteria** are present on **skin**?

Every square inch (6.4 square centimeters) of skin contains approximately 32 million bacteria. Collectively, there are some 100 billion bacteria on the average human body, most of them being harmless.

SKIN STRUCTURE

What are the various **layers** of the **skin**?

Skin is a tissue membrane which consists of layers of epithelial and connective tissues. The outer layer of the skin's epithelial tissue is the epidermis and the inner layer of connective tissue is the dermis. A basement membrane that is anchored to the dermis separates these two layers. The epidermis and dermis rest on a supportive layer of connective tissue and fat cells called the hypodermis. This supportive layer is flexible and allows the skin to move and bend while the fat cells cushion against injury and excessive heat loss.

What causes **warts**?

Warts, which are noncancerous masses produced by uncontrolled growth of epithelial skin cells, are caused by the human papillomavirus. A wart can only be removed by killing the basal cells that harbor the virus. This can be accomplished by cutting away that piece of skin, destroying it by freezing, or killing it with chemicals.

What are the two types of **specialized cells** in the **epidermis**?

The more numerous cells are called keratinocytes which produce a tough, fibrous, waterproof protein called keratin. Keratinization is the process in which cells form fibrils of keratin and harden. Over most of the body, keratinization is minimal, but the palms of the hands and the soles of the feet normally have a thick, outer layer of dead, keratinized cells. Melanocytes, which are less numerous than keratinocytes in the epidermis, produce a pigment called melanin. Melanin ranges in color from yellow to brown to black and determines skin color.

How **common** are **melanocytes** in the **skin**?

The average square inch (6.4 square centimeters) of skin contains 60,000 melanocytes.

What determines **skin color**?

Three factors contribute to skin color: 1) the amount and kind (yellow, reddish brown, or black) of melanin in the epidermis; 2) the amount of carotene (yellow) in the epidermis and subcutaneous tissue; and 3) the amount of oxygen bound to hemoglobin (red blood cell pigment) in the dermal blood cells. Skin color is genetically determined, for the most part. Differences in skin color result not from the number of melanocytes an individual has, but rather from the amount of melanin produced by the melanocytes and the size and distribution of the pigment granules. Although darker-skinned people have slightly more melanocytes than those who are light-skinned, the distribution of melanin in the higher levels of the epidermis contributes to their skin color.

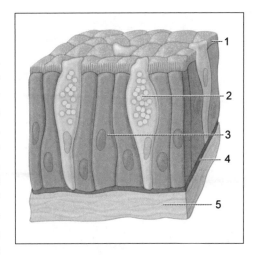

Simple ciliated columnar epithelium. 1 = Microvilli; 2 = Goblet cell; 3 = Absorptive cell; 4 = Basement membrane; 5 = Connective tissue. (From Premkumar K. *The Massage Connection Anatomy and Physiology*. Baltimore: Lippincott, Williams & Wilkins, 2004.)

How does **blood oxygen** level affect **skin color**?

When the blood is well oxygenated, the red blood pigment gives the skin of light-colored individuals a pinkish appearance. When the blood is not well oxygenated, the blood pigment is a darker red, which gives the skin a bluish appearance.

What is **albinism**?

A genetic trait characterized by the lack of ability to produce melanin causes the condition known as albinism. Individuals with this disorder not only lack pigment in the skin, but also in their hair and eyes.

Are **freckles dangerous**?

Freckles, those tan or brown spots on the skin, are small areas of increased skin pigment or melanin. There is a genetic tendency to develop freckles, and parents with freckles often pass this trait down to their children. Freckles usually occur on the face, arms, and other parts of the body that are exposed to the sun. Freckles themselves pose no health risks, but individuals who freckle easily are at an increased risk for skin cancer.

What are **age spots**?

Age spots—also known as sunspots, liver spots, or lentigines—are caused by long term exposure to the sun. Age spots are flat, irregular, brown discolorations of the

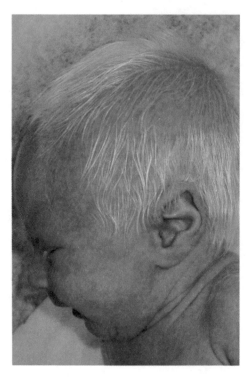

This infant was born with a form of albinism that prevents the formation of skin pigmentation. (From *Stedman's Medical Dictionary*. 27th Ed. Baltimore: Lippincott, Williams & Wilkins, 2000.)

skin that usually occur on the hands, neck, and face of people older than forty. They are not harmful and are not cancerous.

How frequently are **epidermal cells replaced**?

The epidermis is avascular, which means it has no blood supply of its own. This explains why a man can shave daily and not bleed even tough he is cutting off several cell layers each time he shaves. New epidermal cells originate from the deepest layer of the epidermis, the stratum germinativum, and are pushed upward daily to become part of the outermost layer, which flakes off steadily. A totally new epidermis is produced every 25 to 30 days.

What are the **two regions** of the **dermis**?

The dermis is composed of dense connective tissue and is the site of blood vessels, nerves, and epidermal appendages. The dermis has two regions: the papillary and reticular layers. The papillary layer has ridges, which produce fingerprints.

When do **fingerprints form**?

At about 13 weeks of gestation, the human fetus has developed outer epidermal ridges that will eventually develop into fingerprints. These become more and more defined, and at about 21 to 24 weeks the ridges have adult morphology.

Do **identical twins** have the same **fingerprints**?

No. Even identical twins have differences in their fingerprints, which, though subtle, can be discerned by experts. Research indicates that fingerprints would not be the same even in the clone of an individual.

Who first used **fingerprints** as a means of **identification**?

It is generally acknowledged that Francis Galton (1822–1911) was the first to classify fingerprints. However, his basic ideas were further developed by Sir Edward Henry (1850–1931), who devised a system based on the pattern of the thumb print. In 1901 in England, Henry established the first fingerprint bureau with Scotland

How did John Dillinger and Roscoe Pitts attempt to change their fingerprints?

John Dillinger (1903–1934) used acid to burn his fingerprints in an attempt to permanently change them by removing the ridge patterns. He failed, and the fingerprints that reappeared were identical to the ones he had tried to change. In a more dramatic attempt to permanently alter his fingerprints, another American criminal named Roscoe Pitts had a plastic surgeon remove the skin from the first joints of his fingers and replace it with skin grafts from his chest. Investigators were able to identify him from his fingerprints and his palm print.

Yard called the Fingerprint Branch. Today, the number of fingerprints in FBI files is 252 million, compared to 810,000 in 1924.

What is **dermatoglyphics**?

Dermatoglyphics, the study of fingerprints, recognizes three basic patterns of fingerprints. They are arches, loops, and whorls. The lines or ridges of an arch run from one side of the finger to the other with an upward curve in the center. In a loop, the ridges begin on one side, loop around the center, and return to the same side. The ridges of a whorl form a circular pattern. Dermatoglyphics is of interest in such diverse fields as medicine, anthropology, and criminology.

Can **fingerprints** be permanently **changed or destroyed**?

An individual's fingerprints remain the same throughout his or her entire life. Minor cuts or abrasions, and some skin diseases such as eczema or psoriasis, may cause temporary disturbances to the fingerprints, but upon healing the fingerprints will return to their original pattern. More serious injuries to the skin that damage the dermis might leave scars that change or disrupt the ridge pattern of the fingerprints, but examining the skin outside the area of damage will reveal the same fingerprint pattern.

How are **fingerprints** used for **computer security**?

Recent technological advances using optical scanners and solid-state readers use software to analyze the geometric pattern of fingerprints and compare it with those of registered, legitimate users of a network system. Less expensive models of these devices have false acceptance rates of less than 25 per million and false rejection rates of less than three percent. Possible applications include using fingerprints instead of passwords for computers, linking to individual bank accounts and automated teller machines, and for credit cards and Internet transactions.

45

What is the **most abundant** type of **cell** in the **dermis**?

The most abundant cells in the dermis are the fibroblasts, which produce various fibers, including tough collagenous fibers and elastic fibers that give the skin toughness and elasticity.

What is the name of the **protein**—the most abundant in the body—that **holds our skin together**?

The protein that holds our skin together is collagen.

What **other structures** are found in the **dermis**?

Some of the structures in the dermis include:

Hair—The hair root is in the dermis layer and the shaft is above the surface of the skin.

Oil glands—Also known as sebaceous glands, these glands secrete an oily substance that moistens and softens skin and hair.

Sweat glands—These help regulate body temperature.

Blood vessels—They are responsible for supplying the epidermis and dermis with nutrients and removing wastes.

Nerve endings—These provide information about the external environment.

How **thick** is **skin**?

The thickness of skin varies, depending on where it is found on the body. Skin averages 0.05 inches (1.3 millimeters) in thickness. The thinnest skin is found in the eyelids and is less than 0.002 inches (0.05 millimeters) thick, while the thickest skin is on the upper back (0.2 inches or 5 millimeters).

What is the **difference** between **thick and thin skin**?

The terms thick and thin refer to the thickness of the epidermis. Most of the body is covered by thin skin, which is 0.003 inches (0.08 millimeters) thick. This skin contains hair follicles, sebaceous glands, and arrector pili muscles. The epidermis

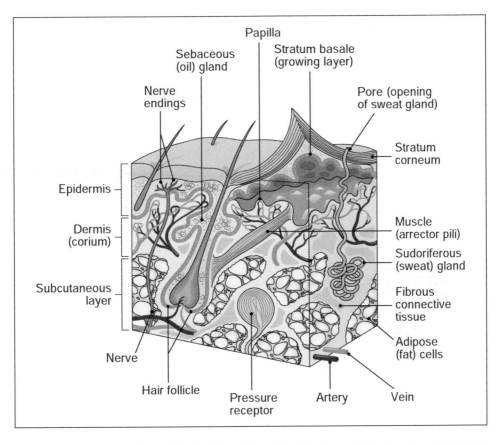

The basic layers and structure of human skin. (From Cohen, B. J., and Wood, D. L. *Memmler's The Human Body in Health and Disease*. 9th Ed. Philadelphia: Lippincott, Williams & Wilkins, 2000.)

in thick skin may be six times thicker than the epidermis that covers the general body surface. Thick skin does not have hair, smooth muscles, or sebaceous glands. Thick skin on the palms of the hands, the fingertips, and soles of the feet may be covered by many layers of keratinized cells that have cornified.

What causes a **blister** to the skin?

The epidermis and dermis are usually firmly cemented together by a basement membrane that is anchored to the dermis. However, a burn or friction due to, for example, rubbing of poorly fitting shoes may cause the epidermis and dermis to separate, resulting in a blister.

What is a **scab**?

A scab is made up of the blood clot and dried tissue fluids that form over a wound. It has an important function in keeping the wound bacteria free while the skin cells underneath divide rapidly to heal the opening. Eventually, the scab will fall off (usually within one or two weeks) and new epithelial tissue will cover the wound.

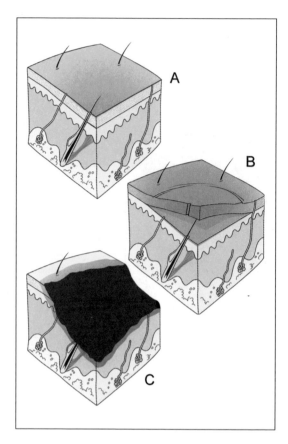

In first-degree burns (A) the skin surface turns pink or red; second-degree burns (B) blister the skin; and third-degree burns (C) severely damage or even burn away parts of the skin. *LifeART image © 2008 Lippincott, Williams & Wilkins.*

Which **layers of skin** are damaged by **burns**?

Burns may be caused by heat generated by radioactive, chemical, or electrical agents. Two factors affect burn severity: the depth of the burn and the extent of the burned area. There are three categories of burns:

First-degree burns—Burns that are red and painful, but not swollen and blistering, such as from a sunburn, and damage only the epidermis.

Second-degree burns—Burns that are red, painful, and blistering, these burns involve injury to the epidermis and the upper region of the dermis.

Third-degree burns—Burns that are severely painful, giving the skin a white or charred appearance; they destroy all layers of the skin, including blood vessels and nerve endings. Skin damaged by third-degree burns does not regenerate. Damage to the skin affects the body's ability to retain fluids.

What are some differences between **cutaneous carcinomas** and **cutaneous melanomas**?

Cutaneous carcinomas (basal cell and squamous cell) are the most common type of skin cancer. They originate from non-pigmented epithelial cells within the deep layer of the epidermis. These cancers usually appear in light-skinned adults who are regularly exposed to sunlight. Cutaneous carcinomas may be flat or raised and develop from hard, dry growths that have reddish bases. This type of carcinoma is slow growing and can usually be completely cured by surgical removal or treatment with radiation.

Melanomas develop from melanocytes and range in color from brown to black and gray to blue. The outline of a malignant melanoma is irregular, rather than smooth, and is often bumpy. Unlike cutaneous carcinomas, melanoma is generally not associated with continued sun exposure. A cutaneous melanoma may arise from normal-appearing skin or from a mole. The lesion grows horizontally but may thicken and grow vertically into the skin, invading deeper tissues. If the melanoma is removed before it invades the deeper tissues, its growth may be arrested. Once it spreads vertically into deeper tissue layers, it is difficult to treat and the survival rate is very low.

How common are **moles** on the body?

Everyone has moles, which are pigmented, fleshy blemishes of the skin. The average person has 10 to 40 moles on his or her skin.

To check for trouble signs indicating melanoma, doctors check spots on the skin for A) assymetry; B) borders; C) colors; and D)diameter. *Anatomical Chart Co.*

What is the **"ABCD" rule** for recognizing **trouble signs** in a **mole or cutaneous melanomas**?

ABCD refers to:

"A" is for *a*symmetry—the two sides of the growth or mole do not match.

"B" is for *b*order irregularity—the border or outline of the growth is not smooth but shows indentations.

"C" is for *c*olor—the pigmented growth contains areas of different colors such as blacks, browns, tans, grays, blues, and reds.

"D" is for *d*iameter—the growth is larger than about 0.25 inches (6.35 millimeters) in diameter, or larger than a pencil eraser.

SKIN FUNCTION

What are the **functions** of the **skin**?

The skin has several different and important functions. It provides protection from both injury (such as abrasion) and dehydration. Since outer skin cells are dead and keratinized, the skin is waterproof, thereby preventing fluid (water) loss. The skin's waterproofing also prevents water from entering the body when a person is immersed. The skin is a barrier against invasion by bacteria and viruses and is involved in the regulation of body temperature. It is the site for the synthesis of an inactive form of vitamin D. In addition, the skin contains receptors that receive the sensations of touch, vibration, pain, and temperature.

How do **skin** cells **synthesize vitamin D**?

Vitamin D is crucial to normal bone growth and development. When ultraviolet (UV) light shines on a lipid present in skin cells, the compound is transformed into

49

vitamin D. People native to equatorial and low-latitude regions of the earth have dark skin pigmentation as a protection against strong, nearly constant exposure to UV radiation. Most people native to countries that exist at higher latitudes—where UV radiation is weaker and less constant—have lighter skin, allowing them to maximize their vitamin D synthesis. During the shorter days of winter, the vitamin D synthesis that occurs in people who live in higher latitudes is limited to small areas of skin exposed to sunlight.

Increased melanin pigmentation, which is present in people native to lower latitudes, reduces the production of vitamin D. Susceptibility to vitamin D deficiency is increased in these populations by the traditional clothing of many cultural groups native to low latitudes, which attempts to cover the body completely to protect the skin from overexposure to UV radiation. Most clothing effectively absorbs irradiation produced by ultraviolet B rays. The dose of ultraviolet light required to stimulate skin synthesis of vitamin D is about six times higher in African Americans than in people of European descent. The presence of darker pigmentation and/or veiling may significantly impair sun-derived vitamin D production, even in sunny regions like Australia.

How do cells become **keratinized**?

The epithelial layer of the skin is continuously replaced. As the replacement cells move closer to the surface of the epidermis, they produce keratin (from the Greek *keras*, meaning "horn"), a tough protein. The transformation of cells into keratin breaks down the cells' nuclei and organelles until they can no longer be distinguished. When the cells' nuclei have broken down, the cells cannot carry out their metabolic functions. By the time the cells reach the superficial layer of the skin, they are dead and composed mostly of keratin.

Do all **stratified squamous epithelial cells** become **keratinized**?

The nuclei and organelles of stratified squamous cells in a moist environment such as the mouth, esophagus, vagina, and cornea do not break down even as the cells reach the superficial layer. This tissue is known as nonkeratinized stratified squamous epithelium.

How fast do **epidermal cells grow** in **tissue culture**?

A piece of epidermis 1.2 square inches (7.7 square centimeters) can be expanded more than 5,000 times within three to four weeks, yielding almost enough skin cells to cover the body surface of an adult human.

How is the **skin** involved in the **regulation of body temperature**?

The skin is one of several organ systems participating in maintaining a core temperature, meaning the temperature near the center of someone's body. Temperature sensors in the skin and internal organs monitor core temperature and transmit sig-

Who was the first individual to culture epidermal cells?

In 1974 Dr. Howard Green of Harvard Medical School discovered the conditions under which the epidermal cells could be grown in the laboratory. The epidermis is especially suitable for growth in laboratory glassware, since it consists of one type of cell—epidermal cells.

nals to the control center located in the hypothalamus, a region of the brain. When the core temperature falls below its set point, the hypothalamus:

1. Sends more nerve impulses to blood vessels in the skin that cause the vessels to narrow, which restricts blood flow to the skin, reducing heat loss.
2. Stimulates the skeletal muscles, causing brief bursts of muscular contraction, known as shivering, which generates heat.

When the core temperature rises above its set point, the hypothalamus:

1. Sends fewer nerve impulses to blood vessels in the skin, causing them to dilate, which increases blood flow to the skin and promotes heat loss.
2. Activates the sweat glands, and when sweat evaporates off the skin surface it carries a large amount of body heat with it.

What is the purpose of **goose bumps**?

The puckering of the skin that takes place when goose-flesh is formed is the result of the contraction of the muscle fibers in the skin. This muscular activity will produce more heat and raises the temperature of the body.

What **skin cells** are involved with the **immune system**?

Keratinocytes, found in the epidermis, assist the immune system by producing hormone-like substances that stimulate the development of certain white blood cells called T lymphocytes. T lymphocytes defend against infection by pathogenic (disease-causing) bacteria and viruses.

NAILS

Why are **nails** part of the **integumentary system**?

Nails, which are modifications of the epidermis, are protective coverings on the ends of the fingers and toes. Each nail consists of a nail plate that overlies a surface of skin called the nail bed. The whitish, thickened, half-moon shaped region (lunula) at the base of the nail plate is the most actively growing region. Nails, like hair, are primarily dead keratinized cells.

51

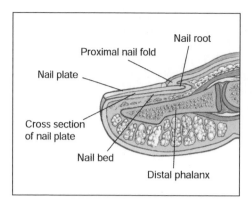

Anatomy of a toenail. (From Bickley, L.S., and Szilagyi, P. *Bates' Guide to Physical Examination and History Taking.* 8th Ed. Philadelphia: Lippincott, Williams & Wilkins, 2003.)

How **fast** do **fingernails grow**?

Healthy nails grow about 0.12 inches (3 millimeters) each month or 1.4 inches (3.5 centimeters) each year. It takes approximately three months for a whole fingernail to be replaced.

Do all **fingernails grow** at the **same rate**?

The thumbnail grows the slowest and the middle nail grows the fastest, because the longer the finger the faster its nail growth.

Do **fingernails** and **toenails grow** at the same **rate**?

Fingernails tend to grow a little faster than toenails.

Are **fingernails and toenails** the same **thickness**?

No. Toenails are approximately twice as thick as fingernails.

HAIR

What are the **three major types** of **hairs**?

The three major types of hairs associated with the integumentary system are:

Vellus hairs—These are the "peach fuzz" hairs found over much of the body.

Terminal hairs—The thick, more deeply pigmented, and sometimes curly hairs that include the hair on the head, eyelashes, eyebrows, and pubic hair.

Intermediate hairs—Those hairs that change in their distribution and include the hair on the arms and legs.

How does **human hair compare** to the hair of **other primates**?

In general, humans have as many hair follicles as gorillas, but the type of hair differs. Gorillas are covered with terminal hairs, while humans are mostly covered with vellus hairs.

How many hairs are on the human body?

On the average human body, there are approximately five million hairs.

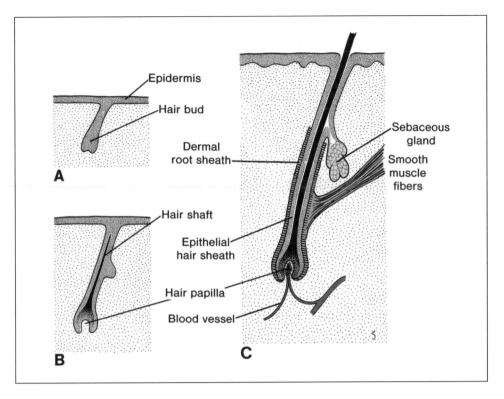

Development of hair in a newborn: A = 4-month-old fetus; B = 6-month-old fetus; C = newborn. (Sadler, T., Ph.D. *Langman's Medical Embryology*. 9th Ed. Image Bank. Baltimore: Lippincott, Williams & Wilkins, 2003.)

How many hairs does the average person have on his or her head?

The amount of hair on the head varies from one individual to another. An average person has about 100,000 hairs on their scalp (blonds 140,000, brunettes 155,000, and redheads only 85,000). Most people shed between 50 to 100 hairs daily.

How fast do eyelashes grow?

Eyelashes are replaced every three months. An individual will grow about 600 complete eyelashes in a lifetime.

How much does hair grow in a year?

Each hair grows about nine inches (23 centimeters) every year.

Does hair grow faster in summer or winter?

During the summertime, human hair grows 10 to 15 percent faster than in the winter. This is because warm weather enhances blood circulation to the skin and scalp, which in turn nourishes hair cells and stimulates growth. In cold weather, when blood is needed to warm internal organs, circulation to the body surface slows and hair cells grow less quickly.

Do the hair and nails continue to grow after death?

No. Between 12 and 18 hours after death, the body begins to dry out. That causes the tips of the fingers and the skin of the face to shrink, creating the illusion that the nails and hair have grown.

Which **hair** on the body **grows the fastest**?

Beards grow more rapidly than any other hair on the body, at a rate of about 5.5 inches (14 centimeters) per year, or about 30 feet (9 meters) in a lifetime.

Is a **hair living** or **dead**?

Hairs are primarily dead, keratinized protein cells produced by the hair bulb. As the cells are pushed farther and farther away from the growing region, the shaft becomes keratinized and dies. The root is enclosed in a sheath, called the hair follicle.

How is **hair color** determined?

Genes determine hair color by directing the type and amount of pigment that epidermal melanocytes produce. If these cells produce an abundance of melanin, the hair is dark. If an intermediate quantity of pigment is produced, the hair is blond. If no pigment is produced, the hair appears white. A mixture of pigmented and unpigmented hair is usually gray. Another pigment, trichosiderin, is found only in red hair.

Why does **hair turn gray**?

The pigment in hair, as well as in the skin, is called melanin. There are two types of melanin: eumelanin, which is dark brown or black, and pheomelanin, which is reddish yellow. Both are made by a type of cell called a melanocyte that resides in the hair bulb and along the bottom of the outer layer of skin, or epidermis. The melanocytes pass this pigment to adjoining epidermal cells called keratinocytes, which produce the protein keratin—hair's chief component. When the keratinocytes eventually die, they retain the melanin. Thus, the pigment that is visible in the hair and in the skin lies in these dead keratinocyte bodies. Gray hair is simply hair with less melanin, and white hair has no melanin at all.

It remains unclear as to how hair loses its pigment. In the early stages of graying, the melanocytes are still present but inactive. Later, they seem to decrease in number. Genes control this lack of deposition of melanin. In some families, many members' hair turns white in their 20s. Generally speaking, among Caucasians 50 percent are gray by age 50. There is, however, wide variation. Premature gray hair is hereditary, but it has also been associated with smoking and vitamin deficiencies. Early onset of gray hair (from birth to puberty) is often associated with various medical syndromes, including dyslexia.

Does shaving make hair coarser?

No. Uncut body hair is tapered and feels softer at the ends. The bristly feeling of hair that has been shaved is the cut end.

What **determines** the **size and shape** of **hairs**?

Hairs are short and stiff in the eyebrows and long and flexible on the head. When the hair shaft is oval, an individual has wavy hair; when it is flat and ribbon-like, the hair is curly or kinky. If the hair shaft is perfectly round, the hair is straight. Humans are born with as many hair follicles as they will ever have, and hairs are among the fastest growing tissues of the body. Each hair follicle on the scalp grows almost 30 feet (9 meters) during an average lifetime.

What are the **arrector pili**?

The arrector pili are tiny, smooth muscles attached to hair follicles, which cause the hair to stand upright when activated.

What accounts for **hairy regions** of the body?

Hormones—particularly androgens—are responsible for the development of such hairy regions as the scalp and, in the adult, the axillary (armpit), chest, and pubic areas.

What is **alopecia**?

Alopecia is the term used to refer to hair loss, which can have many causes. Male pattern baldness, or androgenic alopecia, is an inherited condition. Alopecia areata is characterized by the sudden onset of patchy hair loss. It is most common among children and young adults and can affect either gender.

Does **patterned baldness** have a **genetic basis**?

Pattern baldness is an example of sex-influenced inheritance. Sex-influenced genes are expressed in both males and females, but pattern baldness is expressed more often in males than in females. Male pattern baldness is the most frequent reason for hair loss in men. The gene acts as an autosomal dominant in males and an autosomal recessive in females. One out of every five men begin balding rapidly in their twenties. Another one out of five will always keep their hair. The others will slowly bald over time. The level of baldness is related to the quantity of testosterone in a man.

Is there any **relationship** between **dieting and hair loss**?

There is some scientific evidence that a successful diet can also trigger hair loss. Weight loss can also cause hair to shed, most likely from a nutritional deficiency.

Changes in levels of vitamins A, B, and D, as well as changes in zinc, magnesium, proteins, and fatty acids, can all trigger episodes of hair loss. Hair loss can also be triggered by stress, surgery, pregnancy, and age-related hormonal changes.

ACCESSORY GLANDS

What are the **two types** of **cutaneous glands**?

The two types of cutaneous glands, both of which are found in the dermis, are sebaceous glands and sweat glands. Sebaceous glands, or oil glands, are found all over the skin, except on the palms of the hands and the soles of the feet. Sebaceous glands produce sebum, a mixture of oily substances and fragmented cells that keeps the skin soft and moist and prevents hair from becoming brittle. A second type of cutaneous gland, the sweat glands, are widely distributed in the skin and are an important part of the body's heat-regulating apparatus.

What is the difference between a **whitehead** and a **blackhead**?

The ducts sebaceous glands usually empty into hair follicles, but some open directly onto the surface of the skin. If the duct of a sebaceous gland becomes blocked by sebum, a whitehead appears on the skin surface. If the accumulated material oxidizes and dries, it darkens, forming a blackhead. If the sebaceous glands become infected, resulting in pimples on the skin, the condition is referred to as acne.

What are the **two types** of **sweat glands**?

The two types of sweat glands, also called sudoriferous glands, are eccrine and apocrine glands. Eccrine glands, which are found all over the body and are far more numerous, produce sweat. Sweat is a clear secretion that is largely comprised of water, some salts such as sodium chloride, urea and uric acid (metabolic wastes), and vitamin C. Apocrine sweat glands are found primarily in the axillary (armpit) and genital areas of the body and usually connect to hair follicles. These sweat glands become active when a person is emotionally upset, frightened, or in pain.

What is the length of sweat glands?

Sweat glands are coiled tubes in the dermis, the middle layer of the skin. If one sweat gland were uncoiled and stretched out, it would be approximately 50 inches (127 centimeters) in length. Collectively, the sweat glands are 2,000 miles (3,218 kilometers) in length.

How many sweat glands are present in the body?

Sweat glands are present on all regions of the skin. There can be as many as 90 glands per square centimeter on the leg, 400 glands per cubic centimeter on the palms and soles, and an even greater number on the fingertips. Collectively, there are over two million sweat glands in the adult human body.

How do the eccrine sweat glands work?

The eccrine sweat glands are an important and efficient part of the mechanism that regulates body temperature. These glands are supplied with nerve endings that cause them to secrete sweat when the external temperature or body temperature is high. When sweat evaporates off the skin surface, changing from a liquid to a gas, it carries large amounts of body heat with it.

How numerous are eccrine sweat glands?

The average square inch (6.45 square centimeters) of skin contains 650 eccrine sweat glands.

On a hot day, how much sweat is lost through the skin?

On a typical hot day it is possible to lose up to 7.4 quarts (7 liters) of body water in the form of sweat evaporating off the skin surface. Our bodies lose at least 1 pint (0.473 liters) of sweat every day, even when we are relatively inactive.

Why is there an odor associated with perspiration?

Perspiration, or sweat, is odorless and sterile until bacteria begin to interact with it, giving it a distinctive smell.

How are mammary glands related to the integumentary system?

Mammary glands are modified sweat glands located within the breasts. Each breast contains 15 to 25 lobes, which are divided into lobules. Each lobule contains many alveoli, where milk is secreted and enters a milk duct, which leads to the nipple. Milk is produced only after childbirth.

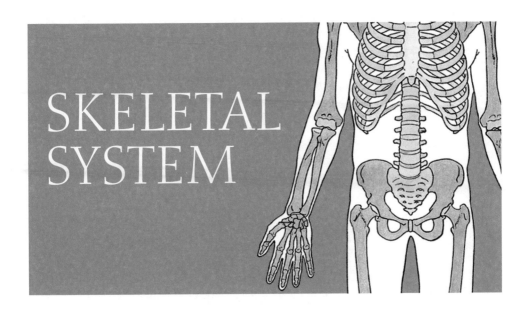

SKELETAL SYSTEM

INTRODUCTION

Who was the **first person** to **study** the **internal structure** of **bone**?

English scientist Clopton Havers (c. 1650–1702) was the first to study the internal structure of bone using a microscope. Havers's discoveries and observations included finding channels that extend along the shafts of the long bones of the arms and legs. These channels, which allow blood vessels to penetrate dense bone, were named the Haversian canals after their discoverer. He also described cartilage and synovia. In addition, he suggested that the periosteum, which surrounds bones, was sensitive to processes occurring within the bone. This observation was not confirmed for 250 years.

What are the **functions** of the **skeletal system**?

The skeletal system has both mechanical and physiological functions. Mechanical functions include support, protection, and movement. The bones of the skeletal system provide the rigid framework that supports the body. Bones also protect internal organs such as the brain, heart, lungs, and organs in the pelvic area. Muscles are anchored to bones and act as levers at the joints, allowing for movement. Physiological functions of the skeletal system include the production of blood cells and the supplying and storing of important minerals.

What are the **major divisions** of the human **skeleton**?

The human skeleton has two major divisions: the axial skeleton and the appendicular skeleton. The axial skeleton includes the bones of the center or axis of the body. The appendicular skeleton consists of the bones of the upper and lower extremities.

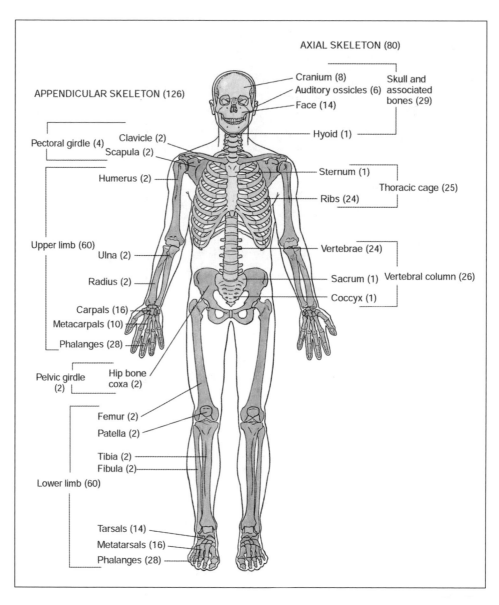

AXIAL SKELETON (80)

APPENDICULAR SKELETON (126)

Cranium (8)
Auditory ossicles (6)
Face (14)

Skull and associated bones (29)

Pectoral girdle (4)
Clavicle (2)
Scapula (2)

Humerus (2)

Hyoid (1)

Sternum (1)
Ribs (24)

Thoracic cage (25)

Upper limb (60)
Ulna (2)

Radius (2)

Vertebrae (24)

Sacrum (1)
Coccyx (1)

Vertebral column (26)

Carpals (16)
Metacarpals (10)

Phalanges (28)

Pelvic girdle (2)
Hip bone coxa (2)

Femur (2)
Patella (2)

Tibia (2)
Fibula (2)

Lower limb (60)

Tarsals (14)
Metatarsals (16)
Phalanges (28)

The human skeleton is divided into two main groups of bones: appendicular and axial. (From Moore, K. L., and Agur, A. M. R. *Essential Clinical Anatomy.* 2nd Ed. Baltimore: Lippincott, Williams & Wilkins.)

How much of the body's **calcium** is **stored in bones**?

Approximately 99 percent of all the calcium found in the body is stored in bones.

Why is **calcium important** to the body?

Bones consist mainly of calcium. Calcium plays an important role as a cofactor for enzyme function, in maintaining cell membranes, in muscle contraction, nervous system functions, and in blood clotting. When the diet does not provide a sufficient amount of calcium, it is released from the bones, and when there is too much calcium in the body, it is stored in the bones.

How **strong** is **bone**?

Bone is one of the strongest materials found in nature. One cubic inch of bone can withstand loads of at least 19,000 pounds (8,626 kilograms), which is approximately the weight of five standard-size pickup trucks. This is roughly four times the strength of concrete. Bone's resistance to load is equal to that of aluminum and light steel. Ounce for ounce, bone is actually stronger than steel and reinforced concrete since steel bars of comparable size would weigh four or five times as much as bone.

Are there **differences** between the **male and female skeletons**?

Several general differences exist between the male and female skeletons. The male skeleton is generally larger and heavier than the female skeleton. The bones of the skull are generally more graceful and less angular in the female skeleton. A female also has a wider, shorter breastbone and slimmer wrists. There are significant differences between the pelvis of a female and a male, which are related to pregnancy and childbirth. The female pelvis is wider and shallower than the male pelvis. Females have an enlarged pelvic outlet and a wider, more circular pelvic inlet. The angle between the pubic bones is much sharper in males, resulting in a more circular, narrower, almost heart-shaped pelvis.

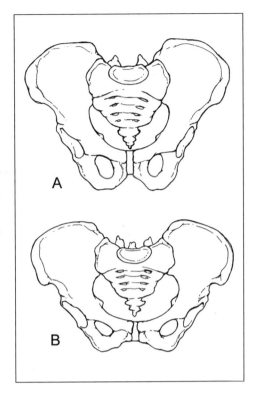

Male (A) and female (B) pelves are easily distinguishable by their shape. (From Oaties, Carol A. *Kinesiology—The Mechanics and Pathomechanics of Human Movement.* Baltimore: Lippincott, Williams & Wilkins, 2004.)

BONE BASICS

How many bones are in the human body?

Babies are born with about 300 to 350 bones, but many of these fuse together between birth and maturity to produce an average adult total of 206. Bone counts vary according to the method used to count them, because a structure may be treated as either multiple bones or as a single bone with multiple parts.

Location of Bones	Number
Skull	22
Ears (pair)	6
Vertebrae	26
Sternum	3
Ribs	24
Throat	1
Pectoral girdle	4
Arms (pair)	60
Hip bones	2
Legs (pair)	58
TOTAL	206

What are the **major types** of **bones**?

There are four major types of bones: long bones, short bones, flat bones, and irregular bones. The name of each type of bone reflects the shape of the bone. Furthermore, the shape of the bone is indicative of its mechanical function. Bones that do not fall into any of these categories are sesamoid bones and accessory bones.

What are the **characteristics** of **long bones**?

The length of a long bone is greater than its width. These bones act as levers that are pulled by contracting muscles. The lever action makes it possible for the body to move. Examples of long bones are the femur (thighbone) and humerus (upper arm bone). Some long bones, such as certain bones in the fingers and toes, are relatively short, but their overall length is still greater than their width.

What is the **longest bone** in the human body?

The femur, or thighbone, is the longest bone in the body. The average femur is 18 inches (45.72 centimeters) long. The longest bone ever recorded was 29.9 inches (75.95 centimeters) long. It was from an 8-foot-tall (2.45 meters) German who died in 1902 in Belgium.

How do **short bones compare** in size to **long bones**?

The terms "long" and "short" are not descriptive of the length of the bones. Short bones have approximately the same dimensions in length, width, and thickness, but they may have an irregular shape. Short bones are almost completely covered with articular surfaces, where one bone moves against another in a joint. The only short bones in the body are the carpal bones in the wrists and tarsal bones in the ankle.

What is the **smallest bone** in the body?

The stirrup (stapes) in the middle ear is the smallest bone in the body. It weighs about 0.0004 ounces (0.011 grams).

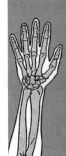

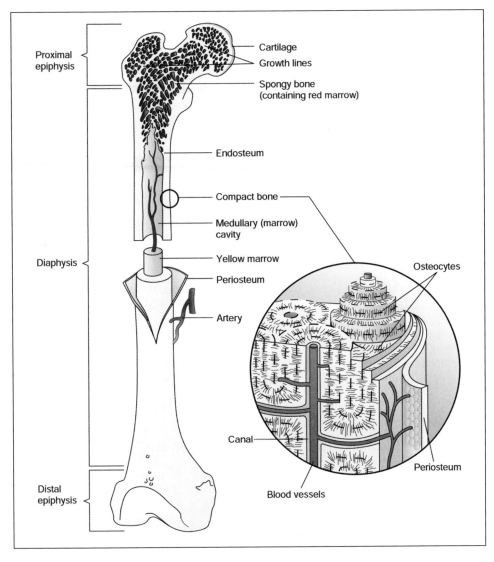

Anatomy of a human bone. (From Smeltzer, S. C., and Bare, B. G. *Textbook of Medical-Surgical Nursing*. 9th Ed. Philadelphia: Lippincott, Williams & Wilkins, 2000.)

Are **flat bones** truly "flat"?

Flat bones are generally thin or curved rather than "flat." Examples of flat bones are the bones that form the cranium of the skull, the sternum or breastbone, the ribs, and scapulae (shoulder blades). The curved shape of most flat bones protects internal organs. The scapula, by contrast, is part of the pectoral (or shoulder) girdle.

What are the **characteristics** of **irregular bones**?

Irregular bones have complex, irregular shapes and do not fit into any other category of bone. Many irregular bones are short, flat, notched, or ridged, with extensions that protrude from their many bone parts. Examples of irregular bones are the spinal vertebrae, many bones of the face and skull, and the hipbones.

What is **unique** about **sesamoid bones**?

Shaped similarly to sesame seeds, sesamoid bones develop inside tendons that pass over a long bone. They are most commonly found in the knees (the patella or kneecap is a sesamoid bone), hands, and feet. Sesamoid bones may form in 26 different locations in the body. However, the number of sesamoid bones varies from individual to individual.

What is the **structure** of a typical **long bone**?

The major parts of a long bone are: epiphysis, epiphyseal plate, metaphysis, diaphysis, medullary cavity, articular cartilage, and periosteum.

Epiphysis—From the Greek, meaning "to grow upon," this spongy bone tissue is spherical in shape and is located at both the distal and proximal end of a long bone.

Epiphyseal plate—This is a layer of hyaline cartilage between the epiphysis and metaphysis. It is the location where bones continue to grow after birth and is therefore often referred to as the epiphyseal growth plate.

Metaphysis—From the Greek *meta,* meaning "between," this is the area of the bone between the epiphysis and diaphysis.

Diaphysis—From the Greek, meaning "to grow between," the diaphysis is the long, cylindrical, hollow shaft of the bone.

Medullary cavity—From the Latin word meaning "marrow," it is the area within the diaphysis and contains fatty (mostly adipose) yellow marrow in adults.

Articular cartilage—A thin layer of hyaline cartilage covering the epiphysis where the bone joins another bone. It helps to reduce friction during joint movement and allows the bones to glide past one another.

Periosteum—From the Greek *peri,* meaning "around," and *osteon,* meaning "bone," it is a white, tough, fibrous membrane that covers the outer surface of the bone whenever it is not covered by articular cartilage. It contains nerves, lymphatic vessels, and blood vessels that provide nutrients to the bone.

Where are **blood cells formed** in the skeletal system?

Hematopoiesis (from the Greek *hemato,* meaning "blood" and *poiein,* meaning "to make"), or red blood formation, occurs in the red bone marrow in adults. Adult red marrow, found in the proximal epiphysis (the ends) of the femur and humerus, some short bones, and in the vertebrae, sternum, ribs, hip bones, and cranium, is the site of production of all red blood cells (erythrocytes), platelets, and certain white blood cells.

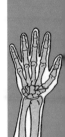

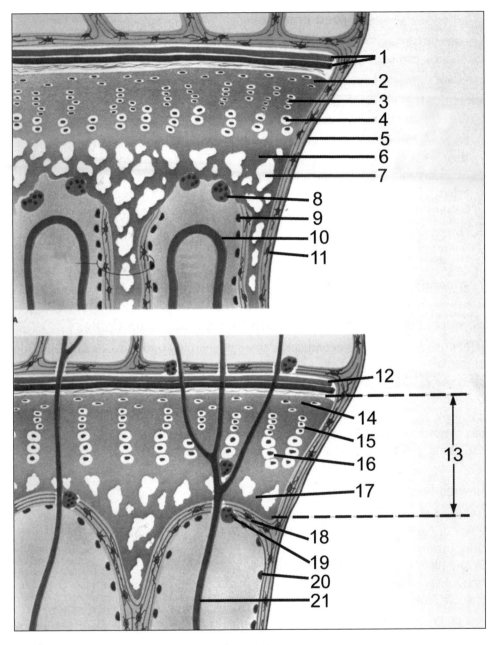

Endochondral ossification is the process where cartilage forms into hard bone. 1 = Epiphyseal artery and vein; 2 = Reserve cartilage; 3 = Proliferating cartilage; 4 = Hypertrophic cartilage; 5 = Periosteum; 6 = Calcified cartilage; 7 = Empty chondrocyte lacuna; 8 = Osteoclast; 9 = Osteoblast; 10 = Vascular loop; 11 = Osteocyte; 12 = Nutritionalartery for epiphyseal plate; 13 = Growth Plate; 14 = Zone of reserve cartilage; 15 = Zone of proliferating cartilage; 16 = Zone of hypertrophied cartilage; 17 = Zone of calcified cartilage; 18 = Transverse bars of bone sealing off plate; 19 = Osteoclast; 20 = Osteoblast; 21 = Metaphyseal artery perforating plate. (From Rubin, E., M.D., and Farber, J. L., M.D. *Pathology.* 3rd Ed. Philadelphia: Lippincott, Williams & Wilkins, 1999.)

What are some **specialized bone cells**?

The four major types of specialized cells in bone are osteogenic cells, osteoblasts, osteocytes, and osteoclasts.

Osteogenic cells—From the Greek *osteo,* meaning "bone," and *genes,* meaning "born," these are cells that are capable of becoming bone-forming cells (osteoblasts) or bone-destroying cells (osteoclasts).

Osteoblasts—From the Greek *osteo* and *blastos,* meaning "bud or growth," they are the cells that form and build bone. Osteoblasts secrete collagen and other organic components needed to build bone tissue. As they surround themselves with matrix materials, they become trapped in their secretions and become osteocytes.

Osteocytes—From the Greek *osteo* and *cyte,* meaning "cell," osteocytes are the main cells in mature bone tissue.

Osteoclasts—From the Greek *osteo* and *klastes,* meaning "break," these cells are multinuclear, huge cells, that are usually found where bone is reabsorbed.

How does **compact bone tissue** differ from **spongy bone tissue**?

Bone tissue is classified as compact or spongy according to the size and distribution of the open spaces in the bone tissue. Compact bone tissue is hard or dense with few open spaces. Compact bone tissue provides protection and support. Most long bones consist of compact bone tissue. In contrast, spongy bone tissue is porous with many open spaces. Spongy bone tissue consists of an irregular latticework of thin needle-like threads of bone called trabeculae (from the Latin *trabs,* meaning "beam"). Most flat, short, and irregular shaped bones are made up of spongy bone tissue.

How do **bones grow**?

Bones form and develop through a process called ossification. There are two types of ossification: intramembranous ossification and endochondral ossification. Intramembranous ossification is the formation of bone directly on or within the fibrous connective tissue. Examples of bone formed through intramembranous ossification are the flat bones of the skull, mandible (lower jaw), and clavicle (collarbone).

Endochondral ossification, from the Greek *endo,* meaning "within," and *khondros,* meaning "cartilage," is the transformation of the cartilage model into bone. Cartilage cells in the epiphyseal plate grow and move into the metaphysis where they are reabsorbed and replaced by bone tissue. Examples of bone formed through endochondral ossification are the long bones, such as the femur and humerus.

What is the **average age** when bone is completely **ossified**?

There is variation among the different bones as to when the epiphyseal plates ossify and the bones fuse. The following table indicates the average age of ossification for different bones.

Bone	Chronological Age of Fusion (years)
Scapula	18–20
Clavicle	23–31
Bones of upper extremity	17–20
Os coxae	18–23
Bones of lower extremity	18–22
Vertebrae	25
Sacrum	23–25
Sternum (body)	23
Sternum (manubrium, xiphoid)	After 30

Do **bones** continue to **grow in diameter** once longitudinal bone growth is completed?

Osteoblasts from the periosteum add new bone tissue around the outer surface of the bone. At the same time, the bone lining the medullary cavity is destroyed by osteoclasts. The combination of new cell growth and erosion of old cells widens the bone without thickening its walls. Bones can also thicken in response to physical changes in the body that may increase the stress or load the bones must support.

What is **remodeling**?

Remodeling is the ongoing replacement of old bone tissue by new bone tissue. In order to maintain homeostasis, bone must be replaced or renewed through the selective resorption of old bone and the simultaneous production of new bone. Approximately 5 to 10 percent of the skeleton is remodeled each year. Thus, every seven years the body grows the equivalent of an entirely new skeleton through remodeling.

How does **exercise** affect **bone** tissue?

Bone adapts to changing stresses and forces. When muscles increase and become more powerful due to exercise, the corresponding bones also become thicker and

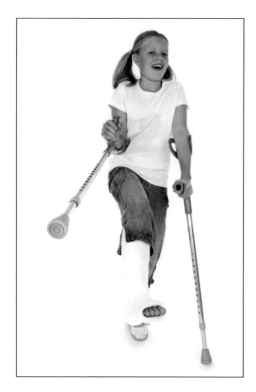

Keeping the bone immobilized is essential for promoting proper mending. © iStockphoto.com/K Fotostudio.

stronger through stimulation of osteoblasts. Regular exercise (especially weight-bearing exercise) maintains normal bone structure. Bones which are not subjected to normal stresses, such as an injured leg immobilized in a cast, quickly degenerate. It is estimated that unstressed bones lose up to a third of their mass after a few weeks. The adaptability of bones allows them to rebuild just as quickly when regular, normal weight-bearing activity is resumed.

How serious is **osteoporosis**?

Osteoporosis (from the Greek *osteo,* meaning "bone," *por,* meaning "passageway," and *osis,* meaning "condition") is a condition that reduces bone mass because the rate of bone resorption is quicker than the rate of bone deposition. The bones become very thin and porous and are easily broken. Osteoporosis is most common in the elderly, who may experience a greater number of broken bones as a result of the mechanical stresses of daily living and not from accidents or other trauma. Generally, osteoporosis is more severe in women, since their bones are thinner and less massive than men's bones. In addition, estrogen helps to maintain bone mass, so the loss of estrogen in women after menopause contributes to more severe osteoporosis.

How does **osteoporosis** differ from **osteomalacia**?

In osteomalacia the bones are weakened and softened from a loss of calcium and phosphorous. The volume of the bone matrix does not change. In osteoporosis, the volume of the bone matrix is reduced, leaving holes in the bones.

AXIAL SKELETON

How many bones are part of the **axial skeleton**?

The adult axial skeleton consists of 80 bones, including the bones of the skull, auditory ossicles (ear bones), hyoid bone, vertebral column, and the bones of the thorax (sternum and ribs).

Which is the only bone that does not touch another bone?

The hyoid bone is the only bone that does not touch another bone. Located above the larynx, it supports the tongue and provides attachment sites for the muscles of the neck and pharynx used in speaking and swallowing. The hyoid is carefully examined when there is a suspicion of strangulation, since it is often fractured from such trauma.

Structure	Number of Bones
Skull	22
Auditory ossicles	6
Hyoid	1
Vertebral column	26
Thorax	25
TOTAL	80

What are the **two sets** of **bones** of the **skull**?

The skull consists of two sets of bones: the cranial and the facial bones. The cranial cavity that encloses and protects the brain consists of the eight cranial bones. There are fourteen facial bones that form the framework of the face. The facial bones also provide support and protect the entrances to the digestive and respiratory systems.

Where is the **vomer bone** located?

The vomer bone (from the Latin, meaning "plowshare") is part of the nasal septum, which divides the nose into left and right halves. A deviated septum, often occurring when the nasal septum has a bend in it, may cause chronic sinusitis and blockage of the smaller nasal cavity. The deviated septum can usually be corrected or improved by surgery.

Do all of the **bones** in the **face and skull move**?

Although the hyoid and ossicles of the ear move, the only bone that can be moved voluntarily and has the greatest range of movement is the mandible. The mandible, the U-shaped bone that forms the jaw and chin, can be raised, lowered, drawn back, pushed forward, and moved from side to side.

What are the functions of the **paranasal sinuses**?

There are four pairs of paranasal sinuses in the bones of the skull located near the nasal cavity. They are lined with mucus. One of their important functions is to act as resonating chambers to produce unique voice sounds. Since they are air-filled, they lighten the weight of the skull bones.

What are the **functions** of the **vertebral column**?

The vertebral column, known also as the spine or backbone, encloses and protects the spinal cord, supports the head, and serves as a point of attachment for the ribs and the muscles of the back. The vertebral column also provides support for the weight of the body, permits movement, and helps the body maintain an erect position.

How **long** is the **vertebral column**?

The average length of the vertebral column is 28 inches (70 centimeters) in males and 24 inches (60 centimeters) in females.

How are the 26 **bones** of the **vertebral column distributed**?

The vertebral column is divided into five regions: cervical, thoracic, lumbar, sacral, and coccygeal.

Region	Number of Bones	Location
Cervical	7	Neck region
Thoracic	12	Chest area
Lumbar	5	Small of the back
Sacral	1 fused bone in adults; 5 individual bones at birth	Below lumbar region
Coccygeal	1 fused bone in adults; 3–5 individual bones at birth	Below sacrum

What is the **general structure** of each **vertebra**?

Each vertebra (from the Latin *vertere,* meaning "something to turn on") consists of a vertebral body, a vertebral arch, and articular processes. The vertebral body is the thick, disc-shaped, front portion of the vertebra that is weight-bearing. The vertebral arch extends backwards from the body of the vertebra. Each vertebral arch has

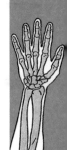

lateral walls called pedicles (from the Latin *pedicle,* meaning "little feet") and a roof formed by flat layers called *laminae* (from the Latin, meaning "thin plates"). The spinal cord passes through the area between the vertebral arch and the vertebral body. Seven vertebral processes (bony projections) extend from the lamina of a vertebra. Some of the processes are attachment sites for muscles. The other four processes form joints with other vertebrae above or below. Intervertebral discs separate each vertebra.

What causes an **intervertebral disc** to **herniate**?

A herniated disc (from the Latin *hernia,* meaning "to bulge" or "stick out") occurs when the soft inner part of an intervertebral disc protrudes through a weakened or torn outer ring and pushes against a spinal nerve. A herniated disc may be the result of an injury or degeneration of the intervertebral

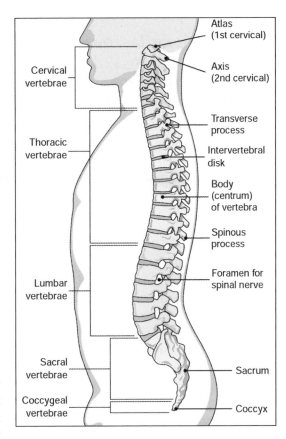

The vertebral column. (From Cohen, B. J., and Wood, D. L. *Memmler's The Human Body in Health and Disease.* 9th Ed. Philadelphia: Lippincott, Williams & Wilkins, 2000.)

joint. Although herniated discs may occur anywhere along the spine, they are most common in the lumbar or sacral regions.

Which **two cervical vertebrae** allow the **head to move**?

The first two cervical vertebrae, C1 and C2, allow the head to move. The first cervical vertebra, the C1 or atlas, articulates with the occipital bone of the skull and makes it possible for a person to nod his or her head. The second cervical vertebra, C2, known as the axis, forms a pivot point for the atlas to move the skull in a side-to-side rotation.

How does the **shape** of the **spine change** from birth to adulthood?

The spine of a newborn infant forms a continuous convex curve from top to bottom. At about three months of age, a concave curve develops in the cervical region as a baby learns to hold up his or her head. When the baby learns to stand during the second half of the first year of life, a concave curve appears in the lumbar region. The thoracic and sacral curves remain the same throughout the life of an individual. These are considered primary curves, since they are present in a fetus and retain their original shape.

What is the **most common abnormal curvature** of the **spine**?

Scoliosis (from the Greek *scolio,* meaning "crookedness") is the most common abnormal curvature of the spine and occurs in the thoracic region or lumbar region or both (thorocolumbar). Individuals with scoliosis have a lateral bending of the vertebral column. In the thoracic area the curves are usually convex to the right, and in the lumbar region curves are usually convex to the left. In most cases of scoliosis, the cause is unknown. Treatment options for scoliosis include observation, bracing, and surgery, depending on the person's age, how much more he or she is likely to grow, the degree and pattern of the curve, and the type of scoliosis.

What are the **causes** of **kyphosis**?

Kyphosis (from the Greek *kypho,* meaning "hunchbacked") is characterized by an exaggeration of the thoracic curve of the vertebral column. In adolescents, kyphosis is often the result of infection or other disturbances of the vertebral epiphysis during growth. In adults, kyphosis may be caused by a degeneration of the intervertebral discs, which results in collapse of the vertebrae. Poor posture may also lead to kyphosis.

Which of the curves of the spine is distorted in **lordosis**?

Lordosis (from the Greek *lord,* meaning "bent backward") is commonly known as "swayback." It is an exaggerated forward curvature of the spine in the lumbar region. The causes of lordosis include poor posture, rickets, tuberculosis of the spine, and obesity. It is not uncommon during the late stages of pregnancy.

What is the **thorax**?

The thorax is a Greek word meaning "breastplate," and it is located the chest. The sternum, twelve costal cartilages, twelve pairs of ribs, and the twelve thoracic vertebrae form the thoracic skeleton, or cage.

What is the **function** of the **thoracic cage**?

The thoracic cage encloses and protects the heart, lungs, and some abdominal organs. It also supports the bones of the shoulder girdle and arm.

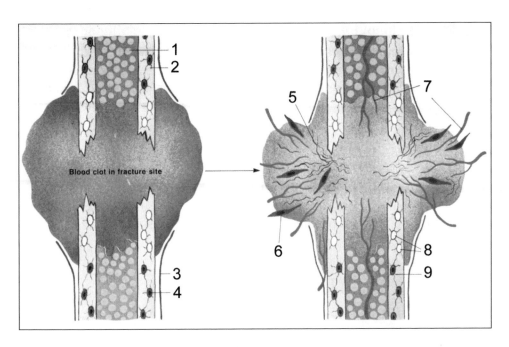
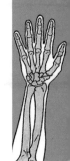

Fibroblasts and collagen work together in the body to heal bone breaks. 1 = Marrow; 2 = Cortical bone; 3 = Periosteum; 4 = Osteocyte; 5 = Collagen; 6 = Fibroblast; 7 = Neovascularization organizing blood clot; 8 = Dead cortex (empty lacunae); 9 = Live cortex. (From Rubin, E., M.D., and Farber, J. L., M.D. *Pathology.* 3rd Ed. Philadelphia: Lippincott, Williams & Wilkins, 1999.)

What are the **parts** of the **sternum**?

The sternum is a flat, narrow bone located in the center of the anterior chest wall. It consists of three parts: manubrium, the body, and the xiphoid process. The manubrium is the upper part and it articulates with the clavicle and first two ribs. The body of the sternum is the largest part of the sternum. It articulates with the second through tenth pairs of ribs. The xiphoid process is the smallest, thinnest, and lowest part of the sternum. Although it does not articulate with any ribs or costal cartilages, several ligaments and muscles are attached to it.

How many **pairs of ribs** does an individual have?

Most individuals have twelve pairs of ribs that form the sides of the thoracic cavity. Approximately five percent of the population is born with at least one extra rib.

How do **true ribs** differ from **false ribs** and **floating ribs**?

The upper seven pairs of ribs are true ribs. These ribs attach directly to the sternum by a strip of hyaline cartilage called costal cartilage. The lower five pairs of ribs are known as false ribs because they either attach indirectly to the sternum or do not attach to the sternum at all. The eighth, ninth, and tenth pairs of ribs attach to each other and then to the cartilage of the seventh pair of ribs. The eleventh and twelfth pairs of ribs are floating ribs because they only attach to the vertebral column and do not attach to the sternum at all.

73

How does the **xiphoid process** change from **infancy** to **adulthood**?

The xiphoid process (from the Greek, meaning "sword-shaped") consists of hyaline cartilage during infancy and childhood. It does not completely ossify until after age 30 and often not until age 40. Proper positioning of the hand during cardiopulmonary resuscitation (CPR) is important to ensure the xiphoid process is not injured.

Which is the **most frequently fractured bone** in the body?

Due to its vulnerable position and its relative thinness, the clavicle is the most frequently fractured bone in the body. Fractured clavicles are caused by either a direct blow or a transmitted force resulting from a fall on the outstretched arm.

APPENDICULAR SKELETON

How many bones are part of the **appendicular skeleton**?

The adult appendicular skeleton consists of 126 bones. It is composed of the bones of the upper and lower extremities, including the pectoral (shoulder) and pelvic girdles, which attach the upper and lower appendages to the axial skeleton.

Structure	Number of Bones
Pectoral (shoulder) girdles	
Clavicle	2
Scapula	2
Upper extremities	
Humerus	2
Ulna	2
Radius	2
Carpals	16
Metacarpals	10
Phalanges	28
Pelvic (hip) girdles	2
Lower extremities	
Femur	2
Fibula	2
Tibia	2
Patella	2
Tarsals	14
Metatarsals	10
Phalanges	28
TOTAL	126

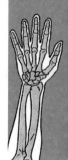

What is the **function** of the **pectoral girdle**?

The pectoral (or shoulder) girdle consists of two bones: the clavicle and scapula. The clavicle articulates with the manubrium of the sternum, providing the only direct connection between the pectoral girdle and the axial skeleton. There is no connection to the vertebral column, allowing for a wide range of movement of the shoulder girdle.

Where is the **funny bone** located?

The funny bone is not a bone but rather a part of the ulnar nerve located at the back of the elbow. A bump or blow to this area can cause a tingling sensation or produce a temporary numbness and paralysis of muscles on the anterior surface of the forearm.

Which **structures** of the body have **more bones** for their size than any other part of the body?

The wrist and the hand have more bones in them for their size than any other part of the body. There are 8 carpals in the wrist between the forearm and the palm of the hand; 5 metacarpal bones that form the palm of the hand between the wrist, thumb, and fingers; and 14 phalanges or finger bones. The presence of many small bones in the wrist and hand with the many movable joints between them makes the human hand highly maneuverable and mobile.

What are the **functions** of the **pelvic girdle**?

The pelvic girdle consists of the two hip bones, also called the coxal bones or ossa coxae. The pelvic girdle provides strong, stable support for the vertebral column, protects the organs of the pelvis, and provides a site for the lower limbs to attach to the axial skeleton.

Which is the **broadest bone** in the body?

The hip bones are the broadest in the body. The hip bones are originally three separate bones in infants: the ilium, ischium, and pubis. These bones fuse together by age 23. The hip bones are united to each other in the front at the pubic symphysis joint and in the back by the sacrum and coccyx.

How does the **pelvic brim** divide the pelvis?

The pelvic brim divides the *pelvis* (from the Latin, meaning "basin") into the false or greater pelvis and the true or lesser pelvis. The part of the pelvis above the pelvic brim is the false pelvis. It does not contain any pelvic organs, except for the urinary bladder, when it is full, and the uterus during pregnancy. The part of the pelvis below the pelvic brim is the true pelvis. The pelvic inlet is the upper opening of the true pelvis, and the pelvic outlet is the lower opening of the true pelvis.

What is the **patella**?

The *patella* (from the Latin, meaning "little plate"), also known as the kneecap, is a small, triangular-shaped, sesamoid bone located within the tendon of the quadriceps femoris, a group of muscles that straighten the knee. The patella protects the knee and gives leverage to the muscles. One of the most common running-related injuries, runner's knee, is essentially an irritation of the cartilage of the kneecap. Normal tracking (gliding) of the kneecap does not occur, since the patella tracks laterally instead of up and down. The increased pressure of abnormal tracking causes pain in this injury.

How is the **shape of the foot** efficient for **supporting weight**?

One of the most efficient types of construction for supporting weight is the arch. The bones of the foot are arranged in two arches to support the weight of the body. The longitudinal arch has two parts, medial and lateral, and extends from the front to the back of the foot. The transverse arch extends across the ball of the foot.

What is the condition of "**fallen arches**"?

The bones of the foot that form the arches are held in place by strong ligaments and tendons. When these ligaments and tendons are weakened by excess weight, postural abnormalities, or genetic predisposition, the height of the medial longitudinal arch of the foot may decrease or "fall," resulting in a condition known as flatfeet.

JOINTS

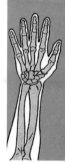

What is a **joint**?

A joint is the place where two adjacent bones, adjacent cartilages, or adjacent bones and cartilages meet. Joints are also called articulations (from the Latin *articulus,* meaning "small joint"). Some joints are very flexible, allowing movement, while others are strong, providing protection of the internal tissues and organs, but do not permit movement.

How are **joints classified**?

There are two classification methods to categorize joints. The structural classification method is based only on the anatomical characteristics of the joint. The functional classification method is based on the type and degree of movement allowed by the joint. Joints are classified both structurally and functionally.

What are the **structural classes** of **joints**?

The two main criteria for the structural classification of joints are the presence or absence of a cavity known as the synovial cavity and the type of tissue that binds the bones together. The three types of structural categories are fibrous, cartilaginous, and synovial.

What are the **three types** of **fibrous joints**?

The fibrous joints are mostly immovable. The three types of fibrous joints are sutures, syndesmoses, and gomphoses. Sutures provide protection for the brain and are only found in the adult skull. They are immovable joints. A syndesmoses joint is a joint where the bones do not touch each other and are held together by fibrous connective tissue. One example of a syndesmoses joint is the distal articulation between the tibia and fibula. A gomphosis joint (from the Greek *gomphos,* meaning "bolt") is composed of a peg and socket. The only gomphoses joints in the human body are the teeth. The roots of the teeth articulate with the sockets of the alveolar processes of the maxillae and mandible.

What are the **functional classes** of **joints**?

The functional classification of joints is determined by the degree and range of movement the joint allows. The three functional categories for joints are synarthrosis, amphiarthrosis, and diarthrosis. A synarthrosis joint (from the Greek *syn,* meaning "together," and *arthrosis,* meaning "articulation") is immovable. An amphiarthrosis joint (from the Greek *amphi,* meaning "on both sides") is slightly movable. A diarthrosis joint (from the Greek *dia,* meaning "between") is a freely movable joint.

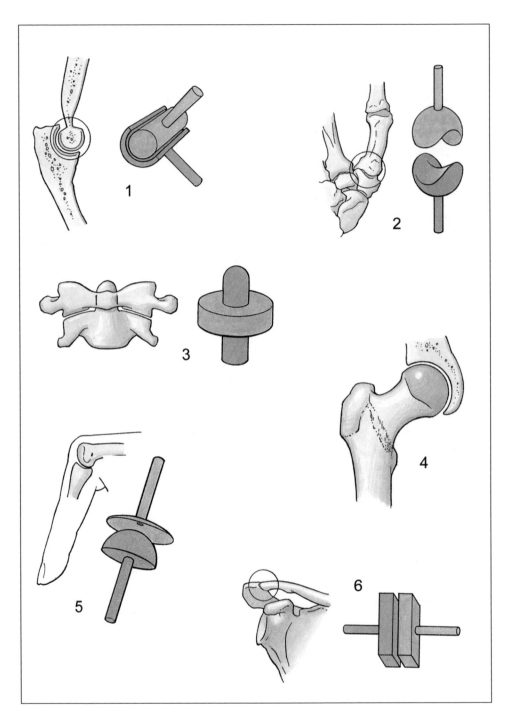

Types of bone joints. 1 = Hinge joint; 2 = Saddle joint; 3 = Pivot joint; 4 = Ball and socket; 5 = Condyloid joint; 6 = Plane joint. (From Moore, K. L., and Agur, A. *Essential Clinical Anatomy*, 2nd Ed. Philadelphia: Lippincott, Williams & Wilkins, 2002.)

Classification of Joints Based on Function

Functional Category	Structural Category	Example
Synarthrosis (immovable joints)	Fibrous	
	Suture	Between bones of adult skull
	Gomphosis	Between teeth and jaw
	Cartilaginous	
	Synchondrosis	Epiphyseal cartilages
Amphiarthrosis (little movement)	Fibrous	
	Syndesmosis	Between the tibia and fibula
	Cartilaginous	Between right and left public bones of pelvis
	Symphysis	Between adjacent vertebral bodies along vertebral column
Diarthrosis (free movement)	Synovial	Elbow, ankle, ribs, wrist, shoulder, hip

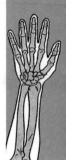

How do the **pubic bones** change **during pregnancy**?

The symphysis pubis, a cartilaginous joint between the two pubic bones, is somewhat relaxed during pregnancy. This allows the mother's hipbones to move in order to accommodate the growing fetus.

Which is the **most common type of joint** found in the body?

Synovial joints are the most common type of joint in the body, permitting the greatest range of movement. Joints of the hip, shoulder, elbow, ankle, and knee are all examples of synovial joints.

What is the **origin** of the **term "synovial fluid"**?

Synovial fluid (from the Greek *syn*, meaning "together," and *ovum*, meaning "egg") is a thick fluid with a consistency similar to the white of an egg. It is a lubricating fluid secreted into joint cavities.

What is the **basic structure** of a **synovial joint**?

The basic structure of a synovial joint consists of a synovial cavity, articular cartilage, a fibrous articular capsule, and ligaments. The synovial cavity (also called joint cavity) is the space between two articulating bones. The articular cartilage covers and protects the bone ends. The articular cartilage also acts as a shock absorber. The articular capsule encloses the joint structure. It consists of an outer layer, the fibrous membrane, and an inner lining, the synovial membrane. Ligaments are fibrous thickenings of the articular capsule that help provide stability.

How many **different types** of **synovial joints** are found in the body?

There are six different types of synovial joints, which are classified based on the shape of their articulating surfaces and the types of joint movements those shapes

A number of reasons have been given for the characteristic "popping" sound associated with someone cracking their knuckles. One reason is that when a joint is contracted, small ligaments or muscles may pull tight and snap across the bony protuberances of the joint. Another possibility is that when the joint is pulled apart, air can pop out from between the bones, creating a vacuum that produces a popping sound. A third reason, discovered by British scientists in 1971, is that when the pressure of the synovial fluid is reduced by the slow articulation of a joint tiny gas bubbles in the fluid may burst, producing the popping sound.

Research has not shown any connection between knuckle cracking and arthritis. One study found that knuckle cracking may be the cause of soft tissue damage to the joint capsule and a decrease in grip strength. The rapid, repeated stretching of the ligaments surrounding the joint is most likely the cause of damage to the soft tissue. Some researchers believe that since the bones of the hand are not fully ossified until approximately age 18, children and teenager who crack their knuckles may deform and enlarge the knuckle bones. However, most researchers believe knuckle cracking does not cause serious joint damage.

permit. The types of synovial joints are gliding or planar joints, hinge joints, pivot joints, condyloid or ellipsoidal joints, saddle joints, and ball-and-socket joints.

Which type of **joint** is most **easily dislocated**?

The ball-and-socket joint is most susceptible to dislocation. Joint dislocation, or luxation (from the Latin *luxare,* meaning "to put out of joint"), occurs when the there is a drastic movement of two bone ends out of their normal end-to-end position. The most frequently dislocated joint is the shoulder joint. Since the socket is shallow and the joint is loose, allowing for the tremendous amount of mobility, there is also the greatest possibility of dislocation.

What are the **four categories** of possible **movements** by synovial joints?

The four basic categories of movements are gliding, angular movements, rotation, and special movements. Each of these groups is defined by the form of motion, the direction of movement, or the relationship of one body part to another during movement.

Gliding movements are generally simple back-and-forth or side-to-side movements. Angular movements include flexion, extension, abduction, adduction, and

What does it mean when someone is "double-jointed"?

Being "double-jointed" (more accurately called joint hypermobility) does not mean having extra joints, but rather having unusually flexible joints, especially of the limbs or fingers. Double-jointed individuals have loose articular capsules.

---

circumduction. In each of these movements there is an increase or decrease in the angle between articulating bones. In rotational movements a bone revolves around its own longitudinal axis. Special movements occur only at certain joints. These movements include elevation, depression, protraction, retraction inversion, eversion, dorsiflexion, plantar flexion, supination, and pronation.

Type of Joint	Type of Movement	Example
Planar	Gliding	Joints between carpals and tarsals
Hinge	Flexion and extension	Elbow, knee, and ankle joints
Pivot	Rotation	Atlantoaxial joint (between first and second vertebrae)
Condyloid	Abduction and adduction	Wrist joint
Saddle	Flexion, extension, metacarpal abduction, adduction	Carpometacarpal joint (between bone of thumb and carpal bone of wrist) circumduction
Ball-and-Socket	Rotation, abduction, adduction, circumduction	Shoulder and hip joints

Which **joint** is the **largest and most complex** joint in the body?

The knee joint (tibiofemoral joint) is the most complex in the human body. It is comprised of three different joints: the medial femoral and medial tibial condyles, the lateral femoral and tibial condyles, and the articulation between the patella and the femur. The knee joint is capable of flexion, extension, and medial and lateral rotation to a certain degree. It is also the joint most vulnerable and susceptible to injury. Common knee injuries are tears to the anterior cruciate ligaments (ACL) and tears to the meniscus or cartilage.

What is **arthritis**?

Arthritis (from the Greek *arthro,* meaning "joint," and *itis,* meaning "inflammation") is a group of diseases that affect synovial joints. Arthritis may originate from an infection, an injury, metabolic problems, or autoimmune disorders. All types of arthritis involve damage to the articular cartilage. Two major categories of arthritis are degenerative diseases and inflammatory diseases.

What is an artificial joint?

Artificial joints are joints designed by engineers to replace diseased or injured joints. Most artificial joints consist of a steel component and a plastic component. For example, an artificial knee joint has three components: the femoral component (made of a highly polished strong metal), the tibial component (made of a durable plastic often held in a metal tray), and the patellar component (also plastic). Artificial joints may be used to replace finger joints, hip joints, or knee joints.

Which is the **most common type** of **arthritis**?

The most common type of arthritis is osteoarthritis. Osteoarthritis is a chronic, degenerative disease most often beginning as part of the aging process. Often referred to as "wear and tear" arthritis because it is the result of life's everyday activities, it is a degradation of the articular cartilage that protects the bones as they move at a joint site. Osteoarthritis usually affects the larger, weight-bearing joints first, such as the hips, knees, and lumbar region of the vertebral column.

How does **rheumatoid arthritis** differ from **osteoarthritis**?

Rheumatoid arthritis is an inflammatory disease mainly characterized by inflammation of the synovial membrane of the joints. The disease may begin with general symptoms of malaise, such as fatigue, low-grade fever, and anemia, before affecting the joints. Unlike osteoarthritis, rheumatoid arthritis usually affects the small joints first, such as those in the fingers, hands, and feet. In the first stage of the disease there is swelling of the synovial lining, causing pain, warmth, stiffness, redness, and swelling around the joint. This is followed by the rapid division and growth of cells, or pannus, which causes the synovium to thicken. In the next stage of the disease, the inflamed cells release enzymes that may digest bone and cartilage, often causing the involved joint to lose its shape and alignment and leading to more pain and loss of movement. This condition is known as fibrous ankylosis (from the Greek *ankulos,* meaning "bent"). In the final stage of the disease the fibrous tissue may become calcified and form a solid fusion of bone, making the joint completely nonfunctional (bony ankylosis).

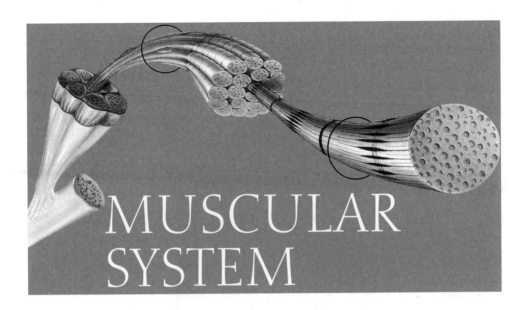

MUSCULAR SYSTEM

INTRODUCTION

What is **myology**?

Myology is the study of the structure and function of muscles.

What are the **functions** of the **muscular system**?

The major functions of the muscular system are:

1. Body movement due to the contraction of skeletal muscles
2. Maintenance of posture also due to skeletal muscles
3. Respiration due to movements of the muscles of the thorax
4. Production of body heat, which is necessary for the maintenance of body temperature, as a byproduct of muscle contraction
5. Communication, such as speaking and writing, which involve skeletal muscles
6. Constriction of organs and vessels, especially smoother muscles that can move solids and liquids in the digestive tract and other secretions, including urine, from organs
7. Heart beat caused by the contraction of cardiac muscle that propels blood to all parts of the body

How **many muscles** are in the **human body**?

There are about 650 muscles in the body, although some authorities believe there are as many as 850 muscles. No exact figure is available because experts disagree about which are separate muscles and which ones branch off larger ones. Also, there is some variability from one person to another, though the general musculature remains the same.

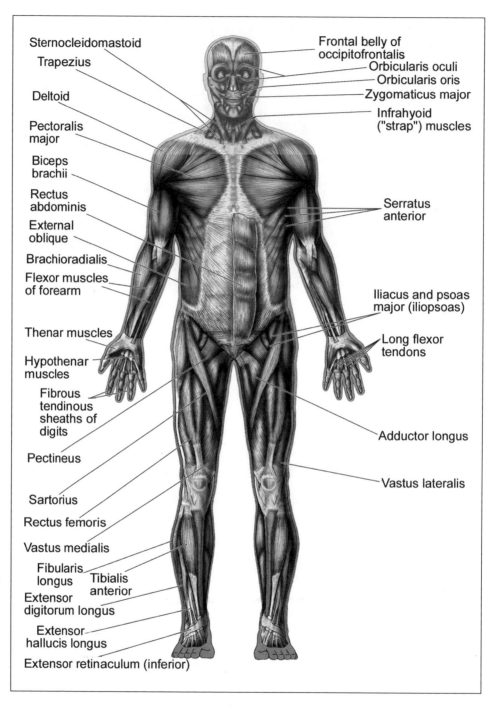

The anterior muscles. (From Moore, K. L., Ph.D., FRSM, FIAC, and Dalley, A.F. II, Ph.D. *Clinical Oriented Anatomy*. 4th Ed. Baltimore: Lippincott, Williams & Wilkins, 1999.)

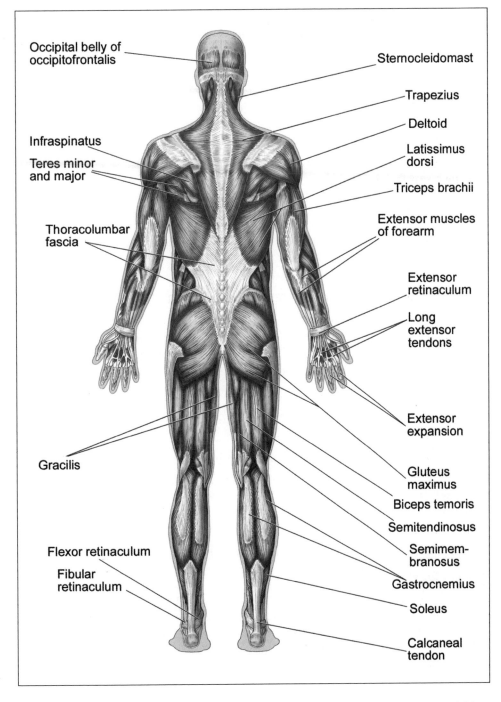

Occipital belly of
occipitofrontalis

Infraspinatus

Teres minor
and major

Thoracolumbar
fascia

Gracilis

Flexor retinaculum

Fibular
retinaculum

Sternocleidomast

Trapezius

Deltoid

Latissimus
dorsi

Triceps brachii

Extensor muscles
of forearm

Extensor
retinaculum

Long
extensor
tendons

Extensor
expansion

Gluteus
maximus

Biceps temoris

Semitendinosus

Semimem-
branosus

Gastrocnemius

Soleus

Calcaneal
tendon

The posterior muscles. (From Moore, K. L., Ph.D., FRSM, FIAC, and Dalley, A.F. II, Ph.D. *Clinical Oriented Anatomy*. 4th Ed.
Baltimore: Lippincott, Williams & Wilkins, 1999.)

85

Does weightlessness in outer space affect skeletal and cardiac muscles?

The effect of weightlessness on humans results in a loss of muscle strength and volume. Similar to bone deterioration, skeletal muscles atrophy as a result of disuse. In space, actions and movement require considerably less exertion because the force of gravity is practically nonexistent. As a result, astronauts' skeletal muscles become deconditioned.

The effects of weightlessness on cardiac (heart) muscles resembles what happens to skeletal muscles in that a weakening can result under such conditions. Just as an athlete will strengthen his or her heart muscles through exercise and make them more efficient as a pump, any reduction in demand will lessen this efficiency.

How **important** and prominent are **muscles cells** in the body?

Muscle cells are found in every organ and tissue in the body and participate in every activity that requires some type of movement. Together, all muscles comprise almost 50 percent of body mass. Nearly 40 percent of body weight in males and almost 32 percent in females is skeletal muscle.

If a man weighs 170 pounds (77 kilograms), **how many pounds of muscle** does he have?

A man who weighs 170 pounds (77 kilograms) has about 81 pounds (37 kilograms) of muscle distributed as 68 pounds (31 kilograms) of skeletal muscle and 13 pounds (6 kilograms) of cardiac and smooth muscle. A woman who weights 120 pounds (54 kilograms) has a total muscle weight of 45 pounds (20 kilograms).

What is the difference between **voluntary** and **involuntary** muscle movements?

Muscle movements that an individual consciously controls are referred to as voluntary. Some examples of voluntary muscle movements would be when an individual walks or picks up an object. Involuntary muscle movements are those that occur without an individual's conscious control. An example of involuntary muscle movement is the pumping action of the heart.

How does **strength training** affect muscles differently than **aerobic training**?

Strength training, such as weight lifting, involves doing exercises that strengthen specific muscles by providing some type of resistance that makes them work harder. It builds more myofibrils (contractile fibers within muscle cells) and increases the size of individual muscle cells (hypertrophy). Strength training builds muscle mass and strength, but it does not increase the number of muscle cells (hyperplasia). In general, the heavier the weight used, the more visible the increase in muscle size. Whereas resistance training strengthens muscles, aerobic training builds endurance. Aerobic training, such as jogging, running, biking, and swimming, involves exercise in which the body increases its oxygen intake to meet the increased demand for oxygen by the muscles. With aerobic training, the number of blood capillaries supplying muscles increases. In addition, the number of mitochondria and the amount of myoglobin available to store oxygen both increase.

Lifting weights is one form of muscle strength training geared toward building muscle mass. © iStockphoto.com/ Michael Monu.

How **quickly** does **muscle strength increase**?

Some muscles gain strength faster than others. In general, large muscles, such as those present in the chest and back, grow faster than smaller ones, including those in the arms and shoulders. Most people can increase their strength between 7 and 40 percent after ten weeks of training each muscle group at least twice a week.

What is an **ecorche**?

An ecorche is a flayed figure, a three-dimensional representation of the human body, usually made of plaster, with the envelope of skin and fat removed. Its intent is to depict the surface muscles with precise anatomical correctness.

Do **creatine supplements** improve **muscular performance**?

Creatine phosphate (CP) is a molecule stored in muscle that yields energy when the creatine splits from the attached phosphate. This energy is used to resynthesize the small amount of ATP (adenosine triphosphate) that is available to the muscle in the initial seconds of high intensity work (think 100-yard dash or a power lift). Because greater amounts of CP in the muscle can potentially allow for those high intensity efforts to be sustained a bit longer or to be performed more effectively, creatine supplementation has become popular within the last 15 years. Some research indicates that such supplementation can improve performance in the short term and in high intensity activities, but for more sustained activities it has little or no effect because of the ATP's great dependence on aerobic metabolism. The long-term effect of such supplementation on the human body is unknown.

ORGANIZATION OF MUSCLES

Which are the **largest** and **smallest muscles** in the human body?

The largest muscle is the gluteus maximus (buttock muscle), which moves the thighbone away from the body and straightens out the hip joint. It is also one of the stronger muscles in the body. The smallest muscle is the stapedius in the middle ear. It is thinner than a thread and 0.05 inches (0.127 centimeters) in length. It activates the stirrup that sends vibrations from the eardrum to the inner ear.

What is the **longest muscle** in the human body?

The longest muscle is the sartorius, which runs from the waist to the knee. Its purpose is to flex the hip and knee.

Which are the **fastest muscles** in the body?

The extraocular muscles, which allow you to move your eye, and the laryngeal muscles associated with vocal folds are the fastest contracting muscles in the body.

What **muscles** act on the skin around the **eyes and eyebrows**?

The occipitofrontalis raises the eyebrows and the orbicularis oculi closes the eyelids and causes "crow's feet" wrinkles in the skin at the lateral corners of the eyes.

How many **muscles** are involved in the **movement** of each **eyeball**?

Six skeletal muscles called the extrinsic eye muscles move the eyeball. They include the superior, inferior, medial, and lateral rectus muscles, as well as the superior and inferior oblique muscles.

How many **muscles** does it take to produce a **smile and a frown**?

Seventeen muscles are used in smiling while the average frown uses 43 muscles.

What muscles are associated with the **lips** and the area surrounding the **mouth**?

The orbicularis oris and buccinator, the kissing muscles, pucker the mouth. The buccinator also flattens the cheeks, as when one whistles or blows a trumpet, and is sometimes called the trumpeter's muscle. Smiling is accomplished primarily by the zygomaticus muscles. Sneering is accomplished by the leavator labii superioris, and frowning or pouting is largely caused by the depressor anguli oris.

How many muscles are involved in chewing food?

Four pairs of muscles are involved in chewing or mastication. These are some of the strongest muscles of the body.

How many muscles are in the human ear?

There are six muscles in the human ear.

What is the function of the corrugator muscle?

Located on the forehead, the corrugator is the muscle that contracts the forehead into wrinkles and pulls the eyebrows together.

What is the triangle of auscultation?

The triangle of auscultation is a small area of the back where three muscles (trapezius, latissimus dorsi, and rhomboideus major) converge. This area is near the scapula and becomes enlarged when a person leans forward with arms folded across the chest. When a physician places a stethoscope on the triangle of auscultation, the sounds of the respiratory organs can be clearly heard.

What are the hamstring muscles?

There are three hamstring muscles, which are located at the back of the thigh. They flex the leg on the thigh, such as when one kneels. They also extend the hip whenever one, for example, sits in a chair. Hamstring injuries are probably the most common muscle injury among runners. Maintaining flexibility and strengthening the muscle helps to prevent injury. Hamstring muscles are also prone to reinjury.

How are muscles named?

Most muscles have names that are descriptive. Muscles are named according to their location, origin and insertion, direction of muscle fibers, size, shape, type of action produced, or other criteria, such as nearby bones.

What **muscles are named** in association with **location in the body**?

The rectus abdominis is a straight muscle located in the abdominal region. Another example is the palmaris longus, which is a long muscle that attaches to connective tissue in the palm of the hand.

What are some examples of **muscles named for size**?

Early anatomists often included in the name of a muscle something about its size, including length. If a muscle were long, its name would likely include the term *longus*. Muscles that were large would have the term *maximus* (Latin for "largest" or "greatest") such as gluteus maximus. Other terms related to size include *brevis* (Latin for "short"), *major* (Latin for "larger"), and *minor* (Latin for "smaller").

What are some examples of **muscles named** on the basis of their **shape**?

Early anatomists often named muscles based upon their resemblance to common shapes. Some muscles named in this manner include the deltoid (a shoulder muscle that gets its name from the Greek *delt*, meaning "triangle" and *oid*, meaning "like," and the trapezeius (another shoulder muscle whose name derives from the Greek word *trapez*, meaning "table".

What are some **muscles named** as a result of the **action produced** by the muscle?

The name of many muscles also includes the type of movement or action that they bring about. The flexors of the wrist and fingers are examples of flexion (from the Latin word *flex*, meaning "to bend") by decreasing the angle at a joint. The adductors of the thigh pulls the limb toward the midline. This type of movement is called adduction (from the Latin word *ad*, meaning "to" or "toward" and *duct*, meaning "lead").

What **muscles** are **named** after the **direction of muscle fibers**?

When looking at a muscle one can often see that it appears to have lines running within it. These lines are made of muscle fibers, and the direction that these fibers run in relation to the midline of the body is often used to provide partial names to different muscles. If the fibers of the muscle are running with or parallel to the body's midline, the term *rectus* is often used to describe that muscle. The translation of rectus from the Latin literally means "straight." Some examples of muscles that have the term rectus in their name include the rectus femoris and rectus abdominis.

If the fibers of the muscle run at an angle to the body's midline, they are said to run obliquely. The term *oblique* is also of Latin origin. Some examples of muscle that have the term oblique associated with their name include the internal and external oblique muscle of the abdominal area.

What **muscles are named** from the location of the muscle's **origin and insertion** on bones?

The first part of the muscle name indicates the origin, while the second part indicates the insertion. For example, the muscle that has its origin on the breast bone and clavicle (collarbone) and that inserts on a breast-shaped process of the skull is termed the sternocleidomastoid (from the Greek words *sterno,* meaning "breast bone," *cleido,* meaning "clavicle," and *mastoid,* meaning "breast shape.")

Are there **other categories** of how **muscles are named**?

There are several other categories of muscle names. One of these is based on the location of the muscle attachment and its association with a particular bone. For example, the temporalis muscle is found covering the temporal bone, while the frontalis muscle is found covering the frontal bone of the skull. Another category deals with the number of origins. Some muscles have multiple origins and the number of origins is often used in the muscle's name. An example is the biceps brachii muscle, which has two heads that attach to two different origins. A final category of muscle names deals with the relation of the muscle to the bone. Not only is a muscle sometimes named because of the bone to which it attached, but the name may be even more detailed to describe where its position is in relation to the bone or body part. Some of the terms and prefixes that describe position are: *supra* (above or over), *infra* (below or beneath), *medialis* (middle), *external* (outer), and *inferior* (underneath). The infraspinatus (an arm muscle connected to the humerus bone) is one example of muscles in this category.

MUSCLE STRUCTURE

What are the **three types of muscle tissue** and their characteristics?

The three types of muscle tissue are skeletal, cardiac, and smooth. The main and most unique characteristic of muscle tissue is its ability to contract, or shorten, making some type of movement possible.

Characteristics of Muscle Tissue

Characteristics	Skeletal Muscle	Smooth Muscle	Cardiac Muscle
Location	Attached to tendons that attach to bones	Found in the walls of blood vessels and in the walls of organs of the digestive, respiratory, urinary, and reproductive tracts and the iris of the eye	Found only in the heart
Function	Movement of the body, maintenance of posture	Control of blood vessel diameter, movement of contents in hollow organs	Pumping of blood

91

Characteristics	Skeletal Muscle	Smooth Muscle	Cardiac Muscle
Cellular characteristics			
Cell shape	Long, cylindrical	Spindle-shaped	Branched
Striations	Present	Absent	Present
Nucleus	Many nuclei	Single nucleus	Single nucleus
Special features	Well-developed transverse tubule system	Lacks transverse tubules	Well-developed transverse tubule system, intercalated discs, separate adjacent cells
Mode of control	Voluntary	Involuntary	Involuntary
Initiation of contraction	Only by a nerve cell	Some contraction always maintained, modifiable by nerves	Autorhythmic (pacemaker cells), modifiable by nerves
Speed of contraction	Fast (0.05 seconds)	Slow (1 to 3 seconds)	Moderate (0.15 seconds)
Sustainability of contraction	Not sustainable	Sustainable indefinitely	Not sustainable
Likelihood of fatigue	Varies widely depending on type of skeletal muscle fiber and work load	Generally does not fatigue	Low—relaxation between contractions reduces the likelihood of fatigue

What are the **four major characteristics** of **skeletal muscle**?

The four major functional characteristics of skeletal muscle are:

Contractility—The ability to shorten which causes movement of the structures to which the muscles are attached.

Excitability—The ability to respond or contract in response to chemical and/or electrical signals.

Extensibility—The capacity to stretch to the normal resting length after contracting

Elasticity—The ability to return to the original resting length after a muscle has been stretched.

Do **cardiac muscle cells** continue to **divide** throughout a person's life?

The vast majority of heart muscle cells are thought to stop dividing by the time a person reaches the age of nine. These cells then pump blood for the rest of a healthy person's life. In people stricken by a heart attack, the cells die and are replaced by scar tissue.

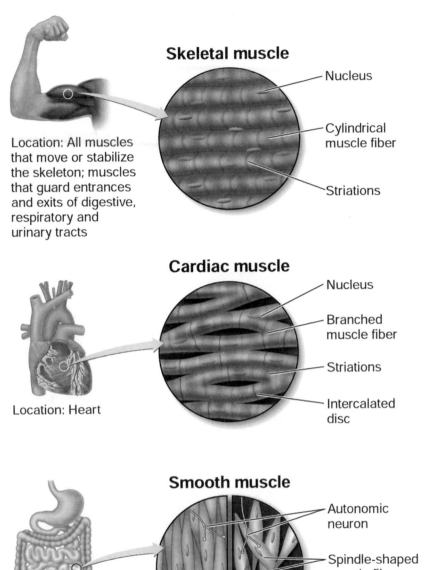

Skeletal muscle

Location: All muscles that move or stabilize the skeleton; muscles that guard entrances and exits of digestive, respiratory and urinary tracts

Nucleus

Cylindrical muscle fiber

Striations

Cardiac muscle

Location: Heart

Nucleus

Branched muscle fiber

Striations

Intercalated disc

Smooth muscle

Other locations: walls of the blood vessels; respiratory tract; urinary and reproductive organs

Autonomic neuron

Spindle-shaped muscle fibers

Nucleus

Visceral (single-unit) smooth muscle tissue

Multiunit smooth muscle tissue

The three types of muscles in the body differ in distinct ways. (From Premkumar K. *The Massage Connection Anatomy and Physiology.* Baltimore: Lippincott, Williams & Wilkins, 2004.)

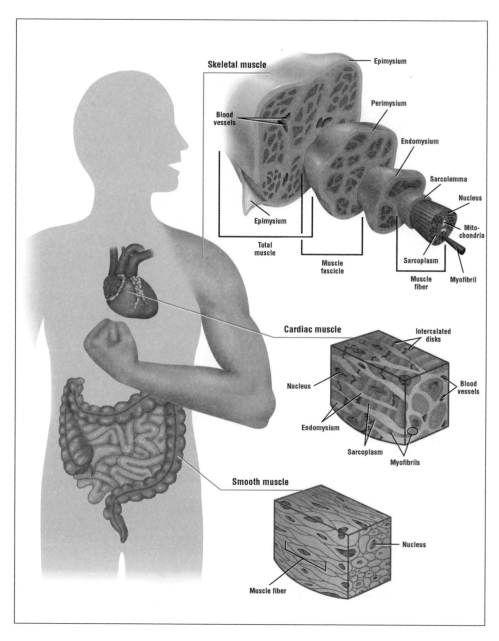

Skeletal, cardiac, and smooth muscles have unique structures. (From Eroschenko, V.P., Ph.D. *di Fiore's Atlas of Histology. With Functional Correlations.* 9th Ed. Baltimore: Lippincott, Williams & Wilkins, 2000.)

What is the basic **contractile unit** of a **muscle cell**?

The basic structural and functional unit of a muscle cell is the sarcomere, which consists of thin filaments of the protein actin and thicker filaments of the protein myosin. The repetition of sarcomeres within the muscle fiber gives the muscle its characteristic striated pattern.

What are **troponin** and **tropomyosin**?

Troponin and tropomyosin are two proteins that are part of the actin filament. Although they do not directly participate in contractions, they help to regulate them.

What is **dystrophin** and how is it related to **muscular dystrophy**?

A muscle cell is packed with filaments of actin and myosin. Less abundant, but no less important, is a protein called dystrophin. It literally holds skeletal muscle cells together by linking actin in the cell to glycoproteins (called dystrophin-associated glycoproteins, or DAGs) that are part of the cell membrane.

What **causes muscular dystrophy**?

Often thought of as a single disorder, muscular dystrophy is a group of genetic diseases characterized by progressive weakness and degeneration of the skeletal muscles that control movement. Missing or abnormal dystrophin causes muscular dystrophies. There are 30 types of muscular dystrophies and they are subdivided by mode of inheritance, age of onset, and clinical features. Discovery of the gene that causes the most common forms of muscular dystrophy took many years because dystrophin compromises only 0.002 percent of the protein in skeletal muscle. There is no specific treatment for muscular dystrophies. Ultimately, fat and connective tissue replace muscle.

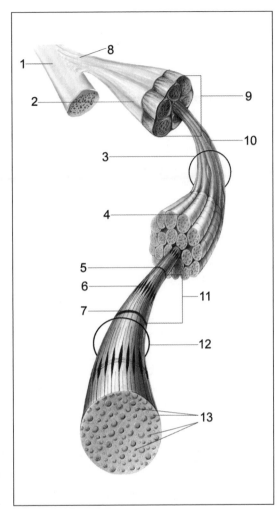

Muscle fiber. 1 = Periosteum: A tough, fibrous connective tissue that covers the surface of bones, rich in sensory nerves, responsible for healing fractures; 2 = Epimysium: Fibrous tissue enveloping the entire muscle and continuous with the tendon; 3 = Fascicle: a group of fibers that have been bound by perimysium; 4 = Endomysium: a delicate connective tissue that surrounds each muscle fiber; 5 = Z-line: boundary of two sarcomeres; 6 = H-zone: consists of stacks of myosin filaments; 7 = Z-line; 8 = Tendon: a dense, fibrous connective tissue that is continuous with the periosteum and attaches muscle to the bone; 9 = Belly: thick contractile portion (or body) of the muscle; 10 = Perimysium: fibrous tissue that extends inward from epimysium, surrounding bundles of muscle. Each bundle bound by perimysium is called a fascicle; 11 = Sarcomere: portion of muscle fibers found between two Z-lines; 12 = Muscle fibers: long, cylindrical, multinucleated cells with striations; 13 = Myosin actin: thick and thin filaments active during muscle contraction. *Anatomical Chart Co.*

What **muscular dystrophies** are **inherited** as X-linked, recessive genetic disorders?

The Duchenne and Becker muscular dystrophies are inherited as X-linked, recessive traits, meaning that only males inherit the condition. Duchenne muscular dystrophy is the most common form of muscular dystrophy. The age of onset is between one and five years of age and affects 1 in every 3,500 males. Progressive muscle weakness continues rapidly, and by 12 years of age affected individuals are typically confined to wheelchairs. Death usually occurs by the age of 20 due to respiratory infection or cardiac failure. The age of onset of Becker muscular dystrophy is between 5 and 25 years, has a slow progression with milder symptoms, and some individuals do have a normal life span.

What is **myoglobin**?

Myoglobin is a pigment synthesized in muscle cells that stores oxygen and is responsible for the reddish brown color of skeletal muscle tissue.

What is the difference between the **origin** and **insertion** of a **muscle**?

The skeleton is a complex set of levers that can be pulled in many different directions by contracting or relaxing skeletal muscles. Most muscles extend from one bone to another and cross at least one joint. One end of a skeletal muscle, the origin, attaches to a bone that remains relatively stationary when the muscle contracts. The other end of the muscle, the insertion, attaches to another bone that will undergo the greatest movement when the muscle contracts. When a muscle contracts, its insertion is pulled toward its origin. The origin is generally closer to the midline of the body and the insertion is farther away.

Do **skeletal muscles** ever have **more than one origin or insertion**?

The biceps brachii, located in the arm, is an example of a muscle that has two origins. This is reflected in the name biceps, which means "two heads." This muscle, which originates on the scapula, extends along the front surface of the humerus and is inserted by means of a tendon on the radium. When the biceps brachii contracts, its insertion is pulled toward its origin, and the forearm flexes at the elbow.

Which type of **lever system** works at the **greatest speed,** but at a **mechanical disadvantage**?

Third-class levers work at the greatest speed, but at a mechanical disadvantage. This arrangement is load-effort-fulcrum. Most skeletal muscles in the body, such as the bicep muscle of the arm, act in third-class lever systems. The biceps lift the distal forearm and anything carried in the hand. The elbow acts as the fulcrum.

What are **tendons**?

Tendons are bundles of white fibrous connective tissue that attach a muscle to a bone. They are similar to ligaments that attach bone to bone. Tendonitis is the

What is the origin of the term "Achilles tendon"?

The Achilles tendon derives its name from a hero in Greek mythology. As a baby, Achilles was dipped into magic water that made him invulnerable to harm everywhere it touched his skin. His mother, however, was holding him by the back of his heel and overlooked submerging his heel under the water. Consequently, his heel was vulnerable and proved to be his undoing. Achilles was shot in the heel with an arrow at the battle of Troy and died. By saying that someone has an "Achilles heel" means that the person has a weak spot that can be attacked.

result of a tendon becoming painfully inflamed and swollen following injury or the repeated stress of an activity. The tendons most commonly affected are those associated with the joint capsules of the shoulder, elbow, and hip, as well as those that move the wrist, hand, thigh, and foot. Rest, ice, and anti-inflammatory medication will often relieve the inflammation.

How does **tennis elbow** differ from **golfer's elbow**?

Tennis elbow and golfer's elbow are both injuries caused by overuse of the muscles of the forearm. Tennis elbow, also called lateral epicondylitis, affects the muscles of the forearm, which attach to the bony prominence on the outside of the elbow. Golfer's elbow, also called medial epicondylitis, affects the muscles of the forearm which attach to the inside of the elbow.

What is unusual about the **calcaneal tendon**?

The gastrocnemius muscle, which is located at the back of the lower leg, forms part of the calf. The distal end of the gastrocnemius joins the calcaneal tendon (also knows as the Achilles tendon) which descends to the heel. The calcaneal tendon is the thickest and strongest human tendon, but it can be partially or completely torn as a result of strenuous athletic activities, including those that involve quick movements and directional changes.

What are the major types of **smooth muscles**?

The two major types of smooth muscles are multiunit and visceral. Multiunit smooth muscles are found in the walls of blood vessels and the iris of the eye. These fibers are separate rather than organized into sheets, and they contract only in response to stimulation by nerve impulses or selected hormones. Visceral smooth muscles, which are more common, are composed of sheets of spindle-shaped cells in close contact with one another. They are found in the walls of organs such as the stomach, intestine, uterus, and urinary bladder. Fibers of visceral muscles can stimulate each other and also display rhythmicity or repeated contractions.

What are the **characteristics** of **smooth muscle** cells?

Smooth muscle cells are elongated with tapering ends containing filaments of actin and myosin in myofibrils that extend the length of the cells. However, the actin and myosin filaments are organized differently than in skeletal muscle and lack striations.

What are **gap junctions**?

Gap junctions are connecting channels made of proteins that permit the movement of ions or water between two adjacent cells. They are commonly found in cardiac and smooth muscle cells.

How do **synergistic muscles** differ from **antagonistic muscles**?

Synergistic muscles are groups of muscles that work together to cause the same movement. Muscles that oppose each other are called antagonistic muscles. Antagonist muscles must oppose the action of an agonist muscle so that movement can occur. For example, when the biceps brachii on the front of the upper arm contracts and shortens (agonist), the triceps brachii must relax and lengthen (antagonist) so that the arm can flex.

Do humans have **dark** and **white muscles** similar to those of a chicken?

A chicken has white wing meat and dark leg meat, and humans are much the same in having dark leg muscles and white arm muscles. These differences in color are due to the use of and demands on the limbs. Dark muscle is specialized for endurance and its color comes from a rich blood supply and high myoglobin content. Endurance in dark muscle is at the expense of speed. Your legs can carry you all day, but they cannot move with the speed of a magician's hand. White muscle specializes in very fast contractions and movements, such as wildly clapping hands or swinging a tennis racquet. White muscle tires quickly because it is less well supplied with blood.

MUSCLE FUNCTION

Who **discovered** how **muscles work**?

Hugh Huxley (1924–) and Andrew Huxley (1917–) (the scientists were unrelated) researched theories regarding muscle contraction. Hugh Huxley was initially a nuclear physicist who entered the field of biology at the end of World War II. He used both X-ray diffraction and electron microscopy to study muscle contraction. Andrew Huxley was a muscle biochemist who obtained data similar to Hugh's, indicating that the contractile proteins thought to be present in muscles are not contractile at all, but rather slide past each other to shorten a muscle. This theory is called the sliding filament theory of muscle contraction.

How do muscle cells work?

Muscle cells—whether the skeletal muscles in the arms or legs, the smooth muscles that line the digestive tract and other organs, or the cardiac muscle cells in the

heart—work by contracting. Skeletal muscle cells are comprised of thousands of contracting units known as sarcomeres. The proteins actin (a thin filament) and myosin (a thick filament) are the major components of the sarcomere. These units perform work by moving structures closer together through space. Sarcomeres in the skeletal muscles pull parts of the body through space relative to each other (e.g., walking or swinging the arms).

To visualize how a sarcomere works:

1. Interlace the fingers of your two hands with the palms facing toward you (represents actin, myosin); fingertips touching
2. Push the fingers together so that the overall length from one thumb to the other is decreased (sarcomere length decreases); allow fingers to slide past each other without bending
3. Any object attached to either thumb would be pulled through space as the fingers move together (sliding filament theory).

What are the **four steps** in the **contraction and relaxation** of a skeletal muscle?

The four key steps are:

1. A skeletal muscle must be activated by a nerve, which releases a neurotransmitting chemical.
2. Nerve activation increases the concentration of calcium in the vicinity of actin and myosin, the contractile proteins.
3. The presence of calcium permits muscle contraction.
4. When a muscle cell is no longer stimulated by a nerve, contraction ends.

What causes **muscle cramps**?

Normally, a muscle at work contracts, tightening to exert a pulling force, then stretches out when the movement is finished or when another muscle exerts force in the opposite direction. But sometimes a muscle contracts with great intensity and stays contracted, refusing to stretch out again. This is a muscle cramp. Muscles contract or lengthen in response to electrical signals from nerves. Minerals such as sodium, calcium, and magnesium, which surround and permeate muscles cells, play a key role in the transmission of these signals. Imbalances in these minerals, as well as certain hormones, body fluids, and chemicals, or a malfunction in the nervous system itself can disrupt the flow of electrical signals and cause a muscle to cramp. Fatigued muscles and cold muscles are more likely to cramp.

What is a **muscle spasm**?

A muscle spasm is a sudden, strong, and painful involuntary contraction. When a muscle is in spasm it feels tight and is described as being in a knot. Muscle spasms occur more frequently in muscles that are overworked or injured. Rest and time resolve most muscle spasms. A "charley horse" is a common name for a muscle spasm in the leg.

What is **writer's cramp**?

Writer's cramp is actually a localized muscle spasm called focal dystonia. It is caused by holding a pen or pencil too long, especially too tightly, and occurs only during handwriting. Relaxing the hand periodically, exercising the hand, holding the pen more loosely, and taking frequent breaks from writing usually solves the problem.

What is a **motor unit**?

A motor unit is all of the muscle cells (approximately 1,500 muscle fibers of skeletal muscle) that are controlled by a single motor neuron.

What is a **neuromuscular junction**?

Each skeletal muscle fiber connects to an axon from a nerve cell called a motor neuron. The connection between the motor neuron and muscle fiber is called a neuromuscular junction.

How does **acetylcholine** interact with **muscle cells**?

Acetylcholine is a neurotransmitter, a chemical substance released by nerve cells that has either an excitatory or inhibitory effect on another excitable cell such as another nerve cell or a muscle cell. In the case of skeletal muscle, acetylcholine excites or activates the muscle cells.

What is the action of **botulinum toxin**?

The bacterium *Clostridium botulinum* produces a poison called botulinum toxin that can prevent the release of acetylcholine from motor neuron axons at neuromuscular junctions, causing botulism, a very serious form of food poisoning. This condition is most likely the result of eating home-processed food that has not been heated enough to kill the bacteria in it or to deactivate the toxin. The endospores of this bacterium are very heat resistant and can withstand several hours of boiling at 212°F (100°C) and ten minutes at 248°F (120°C). Botulinum toxin blocks stimulation of muscle fibers, paralyzing muscles, including those responsible for breathing. Without prompt medical treatment, the fatality rate for botulism is very high.

What is **myasthenia gravis**?

Myasthenia gravis, which usually begins in the face, is a muscular weakness not accompanied by atrophy. It is a chronic, progressive autoimmune disease resulting from the destruction of acetylcholine receptors in the neuromuscular junction. Abnormal antibodies that bind to and destroy acetylcholine receptors can be identified in many people who have myasthenia gravis. Because of the decrease in the number of acetylcholine receptors, the efficiency of neuronal stimulation of muscle fibers decreases, and the muscle is weaker as a result.

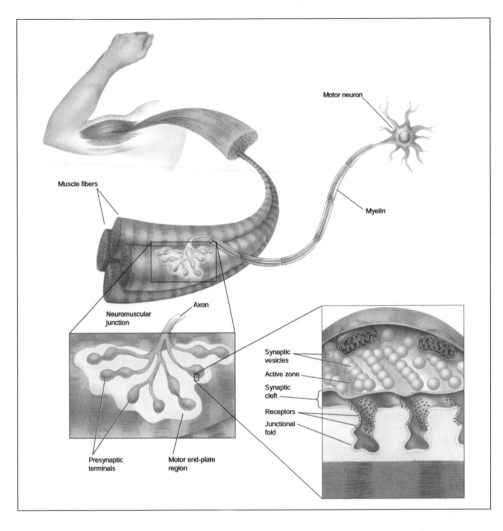

The neuromuscular junction is where nerves connect to muscle. (Bear, Mark F., Connors, Barry W., and Paradiso, Michael A. *Neuroscience: Exploring the Brain.* 2nd Ed. Philadelphia: Lippincott, Williams & Wilkins, 2001.)

Do all **muscle cells work** the **same way**?

Although all muscles work by contracting, not all muscle types have sarcomeres, the muscle contraction units. Cardiac muscle cells have sarcomeres but use different support structures during contraction than those found in skeletal muscles. Smooth muscle cells do not use sarcomeres at all.

How do **muscle cells** use **calcium**?

Calcium ions are stored inside muscle cells. The calcium ions are released from storage when a muscle cell gets a signal to induce contraction, which initiates the movement of the contractile proteins within muscles. When calcium concentrations fall, muscle contractions stop.

101

What **sources** do muscle cells use for **energy**?

Muscle cells use a variety of energy sources to power their contractions. For quick energy, the cells utilize their stores of ATP (adenosine triphosphate) and creatine phosphate, which is another phosphate-containing compound. These stored molecules are usually depleted within the first twenty seconds of activity. The cells then switch to other sources, most notably glycogen, a carbohydrate that is made of glucose molecules strung together in long-branching chains.

What **energy sources** are available for muscles to **contract and relax**?

Muscle contraction requires significant amounts of energy. Like most cells, muscle cells use ATP (adenosine triphosphate) as the energy source. In the presence of calcium ions, myosin acts as an enzyme, splitting ATP into ADP (adenosine diphosphate) and inorganic phosphate and releasing energy to do work. Muscle cells store only enough ATP for about ten seconds worth of activity. Once this is used up, the cells must produce more ATP from other energy sources, including creatine phosphate, glycogen, glucose, and fatty acids.

What is the **all-or-none response** in muscle cells?

According to the all-or-none response, muscle cells are completely under the control of their motor neuron. Muscle cells never contract on their own. A skeletal muscle does not contract partially. If it contracts, it contracts fully.

How is **shivering** related to the muscular system?

Shivering is the body's natural way of keeping warm and can actually serve as a life saver in extreme cold. Shivering produces heat by forcing skeletal muscles to contract and relax rapidly. Heat is produced as a byproduct when muscles metabolize ATP (adenosine triphosphate) for contractions. Approximately 80 percent of the muscle

> ## Why does a runner continue to breathe heavily after completing a ten mile run with a sprint at the end?
>
> **D**uring a ten mile run, aerobic metabolism is the primary source of ATP (adenosine triphosphate) production for muscle contraction. Anaerobic metabolism provides the short (15 to 20 seconds) burst of energy for the sprint to the finish. After the race, aerobic metabolism is elevated for some time to repay the oxygen debt, causing the heavy breathing after the race.

energy used in this process is turned into body heat. One study has shown that warmth from external sources such as blankets and hot water bottles can actually be harmful in some cases of hypothermia because the shivering reflex is shut down.

What is **muscle fatigue**?

Muscle fatigue results from strenuously exercising a muscle for a prolonged period of time. The muscle may lose its ability to contract due to interruption in the muscle's blood supply (and therefore an interruption in the oxygen supply) or the lack of acetylcholine in motor neuron axons. However, muscle fatigue is most commonly associated with the accumulation of lactic acid in the muscle as a result of anaerobic respiration. During vigorous exercise, the circulatory system cannot supply oxygen to muscle fibers quickly enough. In the absence of oxygen, the muscle cells begin to produce lactic acid, which accumulates in the muscle. The lactic acid buildup lowers pH, and as a result muscle fibers no longer respond to stimulation.

Are **pulled muscles** a result of muscle fatigue?

Muscle soreness that develops approximately 24 hours after exercise is the result of microtrauma to the muscle fibers. Pulled muscles, frequently called torn muscles, result from stretching a muscle too far, causing some of the fibers to tear apart. Internal bleeding, swelling, and pain often accompany a pulled muscle.

What is **oxygen debt**?

During rest or moderate exercise, muscles receive enough oxygen to respire aerobically. During strenuous exercise, oxygen deficiency may cause lactic acid to accumulate. Oxygen debt is the amount of oxygen required to convert accumulated lactic acid to glucose and to restore supplies of ATP (adenosine triphosphate) and creatine phosphate.

Does **muscle** regularly **convert to fat** when a person stops **exercising**?

When a person stops exercising regularly, muscles begin to atrophy and fat cells may begin to expand. This process gives the appearance of muscle converting to fat, but the number of muscle cells remains the same.

Are **smooth muscle contractions** the same as **skeletal muscle contractions**?

There are similarities, as well as differences, in comparing smooth and skeletal muscle contractions. Both types of muscles include reactions involving actin and myosin, both are triggered by membrane impulses and an increase in intracellular calcium ions, and both use energy from ATP (adenosine triphosphate). One difference between smooth and skeletal muscle contractions is that smooth muscle is slower to contract and to relax than skeletal muscle. Smooth muscle can maintain a forceful contraction longer with a set amount of ATP. In addition, smooth muscle fibers can change length without changing tautness (as when the stomach is full), while this does not occur in skeletal muscles.

What is **fibrosis**?

Fibrosis is a process in which increasing amounts of fibrous connective tissue develop in skeletal muscle. This makes the muscle less flexible and the collagen fibers can restrict movement and circulation.

What is the **effect of aging** on the **muscular system**?

As the body ages, there is a general reduction in the size and power of all muscle tissues. In general, skeletal muscle fibers become smaller in diameter, reflecting a decrease in the number of myofibrils as well as less ATP (adenosine triphosphate), glycogen reserves, and myoglobin. In addition, skeletal muscles become smaller and less elastic. The tolerance for exercise decreases as does the ability to recover from muscular injuries. Much of the decrease in muscle strength associated with aging is due to decreased activity. Strength training among older adults can help to slow such losses.

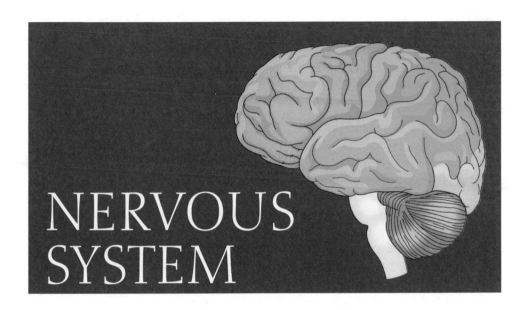

NERVOUS SYSTEM

INTRODUCTION

What are the **functions** of the **nervous system**?

The nervous system is one of the major regulatory systems of the body maintaining homeostasis. Its functions are to: 1) monitor the body's internal and external environments; 2) integrate sensory information; and 3) direct or coordinate the responses of other organ systems to the sensory input.

What are the **two subsystems** of the **nervous system**?

The nervous system is divided into the central nervous system and the peripheral nervous system. The central nervous system consists of the brain and spinal cord, while the peripheral nervous system consists of all the nerve tissue in the body, excluding the brain and spinal cord. Communication between the central nervous system and the rest of the body is via the peripheral nervous system. Specialized cells of the peripheral nervous system allow communication between the two systems.

Are **cells** in the nervous system **replaced** during an individual's lifetime?

Neurons have a very limited capacity for regeneration. In general, they neither replicate themselves nor repair themselves. Axons and dendrites in the peripheral nervous system may undergo repair if the cell body is intact and if the Schwann cells are functional. In the central nervous system, however, a damaged or cut axon is usually not repaired even when the cell body is intact and undamaged. Scientists have discovered recently that there are a few small concentrations of neuronal stem cells that remain in adults that can produce a limited number of new neurons.

105

What are the **two types** of **cells** found in the **peripheral nervous system**?

The peripheral nervous system consists of afferent or sensory neurons and efferent or motor neurons. The afferent nerve cells (from the Latin, *ad,* meaning "toward," and *ferre,* meaning "to bring") carry sensory information from the peripheral to the central nervous system. They have their cell bodies in ganglia and send a process into the central nervous system. The efferent nerve cells (from the Latin *ex,* meaning "away from," and *ferre,* meaning "to bring") carry information away from the central nervous system to the effectors (muscles and tissues). They have cell bodies in the central nervous system and send axons into the periphery.

What are the **divisions** of the **peripheral nervous system**?

The peripheral nervous system is divided into the somatic nervous system and autonomic nervous system. The somatic nervous system has afferent and efferent divisions to receive and process sensory input from the skin, voluntary skeletal (striated) muscles, tendons, joints, eyes, tongue, nose, and ears. The autonomic, or visceral, nervous system innervates smooth muscle and glands.

What are the **divisions** of the **autonomic nervous system**?

The autonomic nervous system is divided into three parts: 1) the sympathetic nervous system; 2) the parasympathetic nervous system; and 3) the enteric nervous system. The parasympathetic and sympathetic nervous systems usually have opposing actions. For example, while the sympathetic nervous system controls the "fight or flight" responses, which increase the heart rate under stress, the parasympathetic nervous system will slow the heart rate. The enteric nervous system consists of nerve cells in the gastrointestinal tract.

What is **amyotrophic lateral sclerosis (ALS)**?

Amyotrophic lateral sclerosis (ALS), also called Lou Gehrig's disease after the New York Yankee baseball player who retired from baseball in 1939 after being diagnosed with ALS, is a fatal neurological disease that attacks the nerve cells (motor neurons) responsible for controlling voluntary muscles. Motor neurons serve as controlling units and vital communication links between the nervous system and the voluntary muscles of the body. Messages from motor neurons in the brain (upper motor neurons) are transmitted to motor neurons in the spinal cord (lower motor neurons) and from them to particular muscles. In individuals with ALS both the upper and

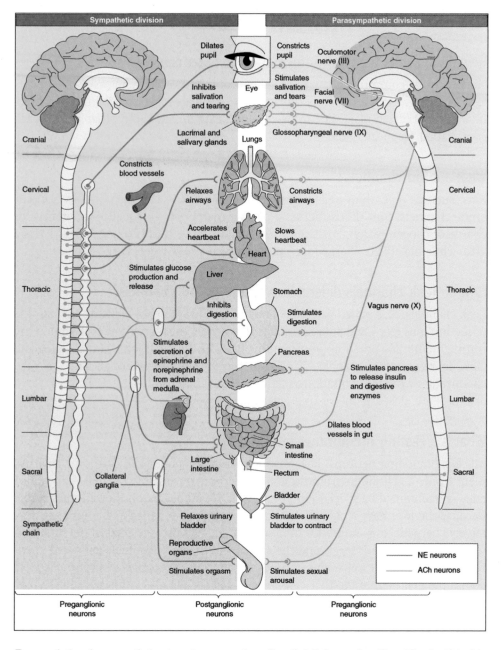

The sympathetic and parasympathetic autonomic nervous systems. (Bear, Mark F., Connors, Barry W., and Paradiso, Michael A. *Neuroscience: Exploring the Brain*. 2nd Ed. Philadelphia: Lippincott, Williams & Wilkins, 2001.)

lower motor neurons degenerate or die and cease to send messages to muscles. Eventually, all muscles under voluntary control are affected and patients lose the strength and ability to move their arms, legs, and other body functions. In the end, even the ability to breathe is affected. The disease does not impair a person's mind, personality, intelligence, or memory.

NEURON FUNCTION

How do **neurons transmit information** to other neurons?

Most neurons communicate with other neurons or muscle by releasing chemicals called neurotransmitters. These transmitters influence receptors on other neurons. In a few specialized places, neurons communicate directly with other neurons via pores called "gap junctions."

What are **ion channels**?

Ions, such as potassium and sodium, pass through the cell membrane via membrane channels. Ions diffuse across a plasma membrane to equalize differences in charge or concentration. Positively charged ions move toward a negatively charged area, while negatively charged ions move toward a positively charged area.

How do **leak channels** differ from **voltage-gated channels**?

Leak channels, also called passive channels, are always open, allowing the passage of sodium ions (Na^+) and potassium ions (K^+) across the membrane to maintain the resting membrane potential of –70 millivolts. Voltage-gated ion channels open and close in response to specific changes in the membrane potential. They may be chemically regulated, voltage regulated, or mechanically regulated. Most gated channels are closed at the resting potential.

What is a **resting membrane potential**?

All cells (including neurons) have resting membrane potentials. The ionic environment inside a cell differs from the ionic environment outside the cell. This difference is maintained by special ion pumps that are embedded in the cell membrane. Because the ions have a charge (cations are positively charged and anions are negatively charged), this difference in ionic content sets up an electrical potential difference between the interior and the exterior of the cell. Excitable cells (e.g., neurons, cardiac muscle cells, and striated muscle cells) also have other ion channels across the cell membrane that can be activated (or gated) by different conditions. For a neuron, this potential difference produced by the ion pumps when the cell is "at rest" is called the "resting membrane potential."

The resting membrane potential of an average neuron is approximately –70 millivolts with respect to the exterior of the cell. This means that the electrical charge on the inside of the plasma membrane measures 0.07 volts less than that on the outside of the plasma membrane.

What is an **action potential**?

An action potential is a series of rapidly occurring events that locally decrease and reverse the membrane potential and then eventually restore it to the resting state. The two phases of an action potential are the depolarizing phase and the repolarizing

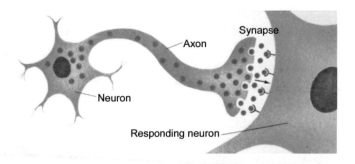

Nerve cells communicate by sending nerve impulses from the axons, across synapses, to the receiving nerve cell. (From Rubin, E., M.D., and Farber, J.L., M.D. *Pathology.* 3rd Ed. Philadelphia: Lippincott, Williams & Wilkins, 1999.)

phase. During the depolarizing phase, the inside of the neuron becomes more positive than the outside, reaching 30 millivolts. During the repolarizing phase, the membrane polarization is restored to its resting state of −70 millivolts. The depolarizing and repolarizing phases of an action potential last approximately one millisecond.

The production of the action potential by the opening of a special ion channel in the cell membrane was first described by Alan Lloyd Hodgkin (1914–1998) and Andrew Fielding Huxley (1917–) in the 1940s. They shared the Nobel Prize in Physiology or Medicine in 1963, along with Sir John Eccles (1903–1997), for their discoveries concerning the ionic mechanisms of the action potential and the excitation and inhibition of the nerve cell membrane.

What is a **synapse**?

A synapse is the location of intercellular communication. Every synapse has components associated with two cells: the presynaptic neuron and the postsynaptic neuron. The presynaptic neuron is the cell that sends the message, while the postsynaptic neuron is the cell that receives the message.

When is a **nerve impulse** generated?

A nerve cell receives many synapses from other neurons (and sometimes from itself). Each time one of these axons conducts an action potential, the presynaptic terminal releases neurotransmitters that can open "chemically gated" ion channels on the postsynaptic neuron (the neuron that receives the terminal). The opening of the ion channels produce local, graded changes in the resting potential of the neuron. If it depolarizes the cell (reduces the potential difference between the inside and outside of the neuron), the small change is called an excitatory postsynaptic potential (or EPSP). If it hyperpolarizes the cell (makes the cell's internal potential more negative with respect its exterior), then it is called an inhibitory postsynaptic potential (or IPSP). All of the EPSPs and IPSPs add to change the membrane potential. A nerve impulse is generated when the membrane potential reaches a critical threshold. This threshold is typically about −55 millivolts. If the neuron does not reach this critical threshold it does not fire an action potential.

Which was the first neurotransmitter to be discovered?

The concept of chemical neurotransmission is generally attributed to Thomas Renton Elliott (1877–1961). As early as 1904, Elliott had published a theory emphasizing the similarity between adrenaline and sympathetic nerve stimulation. It was not until 1921, however, that Otto Loewi (1873–1961) demonstrated experimentally that the transmitter substance at the parasympathetic nerve endings (*Vagusstoff*) is acetylcholine and that a substance closely related to adrenaline played a corresponding role at the sympathetic nerve endings. Loewi shared the Nobel Prize in Physiology or Medicine in 1936 with Sir Henry Hallett Dale (1875–1968) for their discoveries concerning the chemical transmission of nerve impulses.

How quickly do **nerve impulses travel**?

Nerve impulses travel at an average of 160 feet/second (50 meters/second). The slowest nerve impulses travel at 2.5 feet/second (0.7 meters/second) in small unmyelinated (uninsulated) fibers. Nerve impulses in large myelinated (insulated) fibers can travel at 395 feet/second (120 meters/second) or faster.

Are all **action potentials** the **same size**?

Each action potential of a nerve cell is the same for that cell. For this reason, an action potential is called an "all-or-none" response.

What are some **major neurotransmitters**?

Scientists have identified at least fifty neurotransmitters in the nervous system, and there may be several dozen more. There are four groups of neurotransmitters: 1) acetylcholine, 2) amino acids, 3) monoamines, and 4) neuropeptides.

Acetylcholine, perhaps one of the best-known neurotransmitters, is the most important neurotransmitter between motor neurons and voluntary muscle contraction. It has an inhibitory effect on heart muscle and excitatory effect on smooth muscles, through the effects on different types of acetylcholine receptors.

Amino acid neurotransmitters include glutamate and asparate. These neurotransmitters are some of the most potent excitatory neurotransmitters in the central nervous system. They are found in the brain.

There are two important groups of monoamines: catecholamines and indoleamines. Catecholamines include norepinephrine and dopamine. Serotonin, believed to be involved in sleep, mood, appetite, and pain, is an indoleamine.

Neuropeptides include somatostatin, endorphins, and enkephalins. Somatostatin is a growth-hormone inhibiting hormone. Endorphins and enkephalins suppress synaptic activity leading to pain sensation.

How do **excitatory neurotransmitters** differ from **inhibitory neurotransmitters**?

Neurotransmitters are classified as excitatory or inhibitory according to their effects on postsynaptic membranes. A neurotransmitter is called excitatory if activation of the receptor causes depolarization of the membrane and promotes action potential generation. A neurotransmitter is called inhibitory if the activation of the receptor causes hyperpolarization and depresses action potential generation.

What are some **drugs and toxins** which affect **acetylcholine (ACh) activity** at synapses?

The chart below explains the various drugs and toxins and their effects on acetylcholine (ACh) activity.

Drug or Toxin	Mechanism	Effects	Examples
Botulinus toxin (produced by *Clostridium botulinum,* a bacteria)	Inhibits and blocks ACh release	Paralyzes voluntary skeletal muscles	Used therapeutically, such as with Botox, in small doses to remove wrinkles
d-tubocurarine	Prevents ACh binding to postsynaptic receptor sites	Paralyzes voluntary muscles	Known as curare; used by certain South American tribes to paralyze their prey
Atropine	Prevents ACh binding to muscarinic postsynaptic receptor sites	Reduces heart rate and smooth muscle activity decreases salivation; dilation of pupils; high doses produces skeletal muscle weakness	Used therapeutically by ophthalmologists to dilate pupils; may be used therapeutically to counteract the effects of anticholinesterase poisoning
Nicotine	Binds to nicotinic ACh receptor sites and stimulates the postsynaptic membrane	Low doses facilitate voluntary muscles; high doses cause paralysis	Active ingredient in cigarette smoke
Black widow spider venom	Release of ACh	Produces intense muscular cramps and spasms	
Neostigmine or phyostigmine	Prevents ACh inactivation by the enzyme cholinesterase	Extreme sustained contraction of skeletal muscles; effects on cardiac muscle, smooth muscle, and glands	Military nerve gases; used as insecticides (malthion) used therapeutically to treat myasthenia gravis by inhibiting acetylcholinesterase, thereby increasing the usable amount of ACh; counteracts overdoses of tubocurarine

111

How do **local anesthetics** block the sensation of pain?

Local anesthetics, such as Novocain and lidocaine, reduce the permeability of the membrane to sodium. Nerve impulses cannot pass through the membrane, and so the stimulation of sensory neurons is prevented. Pain signals do not reach the central nervous system.

What is **epilepsy**?

Epilepsy is a brain disorder in which clusters of neurons in the brain sometimes signal abnormally. During an epileptic seizure, neurons may fire as many as 500 times a second, much faster than the normal rate of about 80 times a second. When the normal pattern of neuronal activity becomes disturbed, strange sensations, emotions and behavior, convulsions, muscle spasms, and loss of consciousness may be experienced.

What are the **causes** of **epilepsy**?

Epilepsy may develop because of an abnormality in brain wiring, an imbalance of neurotransmitters, or some combination of these factors. Researchers believe that some people with epilepsy have an abnormally high level of excitatory neurotransmitters that increase neuronal activity, while others have an abnormally low level of inhibitory neurotransmitters that decrease neuronal activity in the brain. Either situation can result in too much neuronal activity and cause epilepsy.

What are the different **types** of **seizures**?

There are more than thirty types of seizures, which are categorized as either focal seizures or generalized seizures. Focal seizures, also called partial seizures, occur in just one part of the brain. They are frequently described by the area of the brain in which they originate (e.g., focal frontal lobe seizures). Two examples of focal seizures are simple focal seizures and complex focal seizures. In simple focal seizures, the person will remain conscious but experience sudden and unusual feelings or sensations, such as unexplainable feelings of joy, anger, sadness, or nausea. He or she also may hear, smell, taste, see, or feel things that are not real. In complex focal seizures, the person has a change in or loss of consciousness. People having a complex focal seizure may display strange, repetitious behaviors such as blinks, twitches, mouth movements, or even walking in a circle. These repetitious movements are called automatisms. Some people with focal seizures may experience seeing auras. These seizures usually last just a few seconds.

Generalized seizures are a result of abnormal neuronal activity on both sides of the brain. These seizures may cause loss of consciousness, falls, or massive muscle spasms. There are many kinds of generalized seizures. Two of the better-known generalized seizures are absence seizures and tonic-clonic seizures. In absence seizures, formerly called petit mal seizures, the person may appear to be staring into space and/or have jerking or twitching muscles. Tonic-clonic seizures, formerly called grand mal seizures, cause a mixture of symptoms, including stiffening of the body and repeated jerks of the arms and/or legs, as well as loss of consciousness.

Which potentially fatal neurotoxin is found in puffer fish?

The Pacific puffer fish contains tetrodotoxin, or TTX, which is found in the liver, gonads, and blood of certain species of the fish. Tetrodotoxin blocks the voltage-regulated sodium ion channels, eliminating the production of action potentials and preventing nerve cell activity. The Japanese consider the puffer fish, called fugu, a delicacy. Specially licensed chefs prepare fugu by carefully removing the potentially toxic organs. Although a mild tingling and sense of intoxication are considered desirable, improper preparation of the fish results in several deaths per year.

Which neurotransmitter is depleted in **Parkinson's disease**?

Parkinson's disease results from a deficiency of the neurotransmitter dopamine in certain brain neurons that regulate motor activity. Parkinson's disease is characterized by stiff posture, tremors, slowness of movement, postural instability, and reduced spontaneity of facial expressions.

There is no cure for Parkinson's disease, but certain medications provide relief from the symptoms by increasing the amount of dopamine in the brain. Patients are usually given levodopa combined with carbidopa. Carbidopa delays the conversion of levodopa into dopamine until it reaches the brain. Nerve cells can use levodopa to make dopamine and replenish the brain's dwindling supply.

When was **Parkinson's disease** first **described**?

Parkinson's disease was first formally described by Dr. James Parkinson (1755–1824), a London physician, in "An Essay on the Shaking Palsy," published in 1817.

CENTRAL NERVOUS SYSTEM

What are the **characteristics** of the **central nervous system**?

The central nervous system (brain and spinal cord) is protected by a bony covering. The skull surrounds the brain and the vertebral column protects the spinal cord.

Which **membranes** cover and **protect** the **brain and spinal cord**?

The meninges (from the Greek *meninx*, meaning "membrane") cover and protect the brain and spinal cord. The meninges have three layers: 1) the dura mater, 2) the arachnoid, and 3) the pia mater. The dura mater is the outermost layer covering the central nervous system. The arachnoid (from the Greek *arachne*, meaning "spider") is a weblike network of collagen and elastic fibers. The innermost layer of the meninges is the pia mater. The pia mater is firmly attached to the neural tissue of

113

the spinal cord and brain. Cerebrospinal fluid fills the space between the pia matter and the arachnoid membrane. Most of the blood vessels that supply blood to the central nervous system are in the pia mater.

What is **meningitis**?

Meningitis is an infection or inflammation of the meninges. Meningitis is most often caused by a bacterial or viral infection, although certain fungal infections and tumors may also cause meningitis. The usual symptoms and signs of meningitis are sudden fever, severe headache, and a stiff neck. In more severe cases, neurological symptoms may include nausea and vomiting, confusion and disorientation, drowsiness, sensitivity to bright light, and poor appetite. Early treatment of bacterial meningitis with antibiotics is important to reduce the risk of dying from the disease.

Can **meningitis** be **prevented**?

The introduction and widespread use of *Hemophilus influenzae* type b and *Streptococcus pneumoniae* conjugated vaccines has dramatically reduced the incidence of meningitis caused by these bacteria. In 2005, the Centers for Disease Control recommended routine vaccination of adolescents and college freshmen with the new meningococcal vaccine, which prevents four types of meningococcal disease caused by the bacteria *Neisseria meningitides.*

What is **gray matter**?

Gray matter consists of neurons and unmyelinated dendrites and axons. In the spinal cord the gray matter is shaped like an "H" around the very small, narrow central canal. It has a grayish appearance in autopsy specimens.

What is **white matter**?

White matter consists of myelinated nerve tissue. Since myelin is white the tissue appears whitish.

What are **demyelinating diseases**?

Demyelinating diseases involve damage to the myelin sheath of neurons in either the peripheral or central nervous system. Multiple sclerosis (MS) is a chronic, potentially debilitating disease that affects the myelin sheath of the central nervous system. The illness is believed to be an autoimmune disease. In MS the body directs antibodies and white blood cells against proteins in the myelin sheath surrounding nerves in the brain and spinal cord. This causes inflammation and injury to the myelin sheath. Demyelination is the term used for a loss of myelin, a substance in the white matter that insulates nerve endings. Myelin helps the nerves receive and interpret messages from the brain at maximum speed. When nerve endings lose this substance, they cannot function properly, leading to patches of scarring, or "sclerosis." The result may be multiple areas of sclerosis. The damage slows or blocks muscle coordination, visual sensation, and other functions that rely on nerve signals.

How is the **myelin sheath** affected in **Guillain-Barré syndrome**?

In the autoimmune disorder known as Guillain-Barré syndrome, the body's immune system attacks part of the peripheral nervous system. The immune system starts to destroy the myelin sheath that surrounds the axons of many peripheral nerves, or even the axons themselves. The loss of the myelin sheath surrounding the axons slows down the transmission of nerve signals. In diseases such as Guillain-Barré; in which the peripheral nerves' myelin sheaths are injured or degraded, the nerves cannot transmit signals efficiently. Consequently, muscles begin to lose their ability to respond to the brain's commands, commands that must be carried through the nerve network. The brain also receives fewer sensory signals from the rest of the body, resulting in an inability to feel textures, heat, pain, and other sensations. Alternatively, the brain may receive inappropriate signals that result in tingling, "crawling-skin," or painful sensations. Because the signals to and from the arms and legs must travel the longest distances, these extremities are most vulnerable to interruption.

What are the **symptoms** of **Guillain-Barré syndrome**?

The first symptoms of this disorder include varying degrees of weakness or tingling sensations in the legs. In many instances the weakness and abnormal sensations spread to the arms and upper body. In severe cases the patient may be almost totally paralyzed since the muscles cannot be used at all. In these cases the disorder is life threatening because it potentially interferes with breathing and, at times, with blood pressure or heart rate. Such a patient is often put on a respirator to assist with breathing and is watched closely for problems such as an abnormal heart beat, infections, blood clots, and high or low blood pressure. Most patients, however, recover from even the most severe cases of Guillain-Barré syndrome, although some continue to have a certain degree of weakness.

How much **cerebrospinal fluid** is in the central nervous system?

The entire central nervous system contains between 3 to 5 ounces (80 to 150 milliliters) of cerebrospinal fluid, a clear, colorless liquid. The choroid plexus produces nearly 17 ounces (500 milliliters) of cerebrospinal fluid per day, effectively replacing the cerebrospinal fluid every eight hours (three times per day). Normally, cerebrospinal fluid flows through the ventricles, exits into cisterns (closed spaces that serve as reservoirs) at the base of the brain, bathes the surfaces of the brain and spinal cord, and then is absorbed into the bloodstream.

What are the **functions** of **cerebrospinal fluid**?

Cerebrospinal fluid has three important, life-sustaining functions: 1) to keep the brain tissue buoyant, acting as a cushion or "shock absorber"; 2) to act as the vehicle for delivering nutrients to the brain and removing waste; and 3) to flow between the cranium and spine to compensate for changes in intracranial blood volume (the amount of blood within the brain).

What **condition** results from **interference** with the circulation of **cerebrospinal fluid**?

Obstructive hydrocephalus, commonly called "water on the brain," results from an imbalance of production, circulation, and reabsorption of cerebrospinal fluid. Since cerebrospinal fluid is being produced continually, once the balance is disrupted the volume of cerebrospinal fluid within the brain will continue to increase. The increased volume of fluid leads to compression and distortion of the brain. Left untreated, the intracranial pressure increases, often causing brain function to deteriorate. In infants, treatment often includes the installation of a shunt to either avoid the site of the blockage or drain the excess cerebrospinal fluid.

THE BRAIN

How **large** is the **brain**?

The brain weighs about 3 pounds (1.4 kilograms). The average brain has a volume of 71 cubic inches (1,200 cubic centimeters). In general, the brain of males averages about 10 percent larger than those of females due to overall differences in average body size. The brain contains approximately 100 billion neurons and 1 trillion neuroglia.

Is **brain size** an **indication** of **intelligence**?

There is no correlation between brain size and intelligence. Individuals with the smallest brains (as small as 46 cubic inches [750 cubic centimeters]) and the largest brains (as large as 128 cubic inches [2,100 cubic centimeters]) have the same functional intelligence.

How does the **size** of the **brain** change from **birth to adulthood**?

Brain cells grow in size and degree of myelination as a child grows from birth to adulthood. Although the number of neurons does not increase after infancy, the number of glial cells does increase. An adult brain is approximately three times as heavy as it was at birth. Between ages 20 and 60, the brain loses approximately 0.033 to 0.10 ounces (1 to 3 grams) a year as neurons die and are not replaced. After age 60 the annual rate of shrinkage increases to 0.10 to 0.143 ounces (3 to 4 grams) per year.

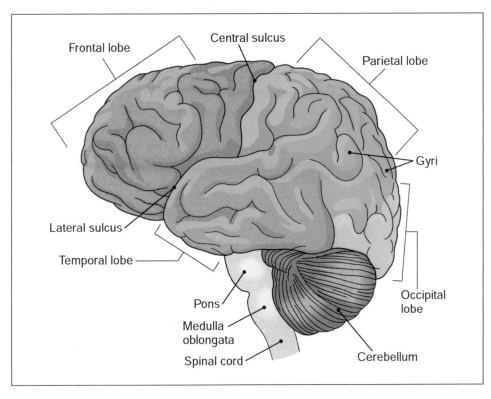

Frontal lobe

Central sulcus

Parietal lobe

Gyri

Lateral sulcus

Temporal lobe

Occipital lobe

Pons

Medulla oblongata

Spinal cord

Cerebellum

The brain is comprised of four main regions: the brain stem (pons and medulla oblongata), cerebellum, diencephalon (includes the thalamus and hypothalamus), and cerebrum, which includes the lobes, sulci, and gyri. (From Cohen, B.J. *Medical Terminology.* 4th Ed. Philadelphia: Lippincott, Williams & Wilkins, 2003.)

What are the **major divisions** of the **brain**?

The brain has four major divisions: 1) brainstem, including the medulla oblongata, pons, and midbrain; 2) cerebellum; 3) cerebrum; and 4) diencephalon. The diencephalon is further divided into the thalamus, hypothalamus, epithalamus, and ventral thalamus or subthalamus.

What are the **functions** of the **major divisions** of the **brain**?

Each area of the brain has a specific function, as seen in the below table.

Brain Area	General Functions
Brainstem	
Medulla oblongata	Relays messages between spinal cord and brain and to cerebrum; center for control and regulation of cardiac, respiratory, and digestive activities
Pons	Relays information from medulla and other areas of the brain; controls certain respiratory functions
Midbrain	Involved with the processing of visual information, including visual reflexes, movement of eyes, focusing of lens, and dilation of pupils

Brain Area	General Functions
Cerebellum	Processing center involved with coordination of movements, balance and equilibrium, posture; processes sensory information used by motor systems
Cerebrum	Center for conscious thought processes and intellectual functions, memory, sensory perception, and emotions
Diencephalon	
Thalamus	Relay and processing center for sensory information
Hypothalamus	Regulates body temperature, water balance, sleep-wake cycles, appetite, emotions, and hormone production

Which **structure connects** the **spinal cord** to the **brain**?

The medulla oblongata connects the spinal cord to the brain. The medulla oblongata regulates autonomic functions, such as heart rate, blood pressure, and digestion, and automatic functions, such as respiratory rhythm. It relays sensory information to the thalamus and to other portions of the brain stem.

Which is the **largest part** of the **brain**?

The largest part of the brain is the cerebrum. The outer surface of the cerebrum is covered with a series of elevated ridges called gyri, and grooves or shallow depressions called sulci. The deepest sulci are called fissures. The cerebrum is divided into the right and left hemispheres. The corpus callosum connects the two halves at their lower midportion. Each hemisphere is divided into four sections called lobes, which have been named for the bones of the skull that cover them. The lobes are identified as the frontal lobe, the parietal lobe, the temporal lobe, and the occipital lobe.

Which **skills** are controlled by the **left cerebral hemisphere** and which are controlled by the **right cerebral hemisphere**?

The left side of the brain controls the right side of the body, as well as spoken and written language, logic, reasoning, and scientific and mathematical abilities. In

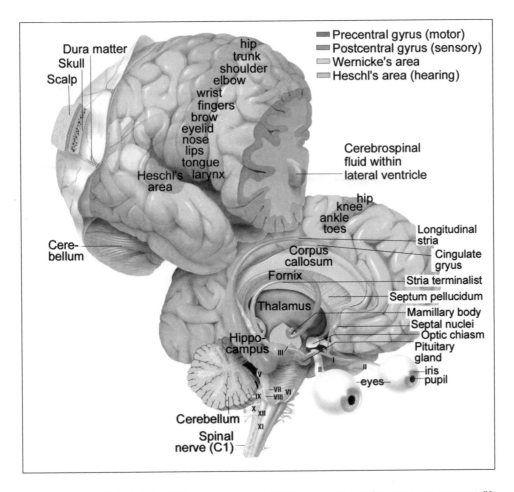

Precentral gyrus (motor)
Postcentral gyrus (sensory)
Wernicke's area
Heschl's area (hearing)

Dura matter
Skull
Scalp

hip
trunk
shoulder
elbow
wrist
fingers
brow
eyelid
nose
lips
tongue
Heschl's larynx
area

Cerebrospinal
fluid within
lateral ventricle

Cere-
bellum

hip
knee
ankle
toes

Corpus
callosum

Fornix

Thalamus

Hippo-
campus

Cerebellum

Spinal
nerve (C1)

Longitudinal
stria

Cingulate
gryus

Stria terminalist

Septum pellucidum

Mamillary body
Septal nuclei
Optic chiasm
Pituitary
gland
iris
pupil

eyes

A detailed look at the brain, including: I) Olfactory nerve—smell; II) Optic nerve—sight; III) Oculomotor—eye movement; IV) Trochlear nerve—eye movement (not illustrated); V) Trigeminal nerve—face (sensory); VI) Abducens nerve—eye movement; VII) Facial nerve—face (motor), taste; VIII) Vestibulocochlear nerve (hearing and balance); IX) Glossopharyngeal nerve—swallowing, taste, sensation; X) Vagus nerve—gastrointestinal tract, swallowing, heart rate, peristalsis; XI) Accessory nerve—shoulder muscles; and XII) Hypoglossal nerve—tongue. *Anatomical Chart Co.*

contrast, the right side of the brain controls the left side of the body and is associated with imagination, spatial perception, recognition of faces, and artistic and musical abilities.

Why is the **blood-brain barrier** important?

The blood-brain barrier is formed by the contacts of special glial cells, called astrocytes, with blood vessels. It is essential for maintaining homeostasis in the brain. In general, only lipid-soluble molecules, such as carbon dioxide, oxygen, steroids, and alcohols, can pass through the blood-brain barrier easily. Water-soluble molecules, such as sodium, potassium, and chloride ions can pass through the blood-brain barrier only with the assistance of specific carrier molecules. Some substances cannot pass through the barrier at all.

119

What is a **concussion**?

A concussion is an injury to the brain caused by a blow or jolt to the head that disrupts the normal functioning of the brain. Concussions are usually not life-threatening. Since the brain is very complex, there is great variation in the signs and symptoms of a concussion. Some people lose consciousness; others never lose consciousness. Some symptoms may appear immediately, while others do not appear for several days or even weeks. Symptoms include:

- Headaches or neck pain that will not go away
- Difficulty with mental tasks such as remembering, concentrating, or making decisions
- Slowness in thinking, speaking, acting, or reading
- Getting lost or easily confused
- Feeling tired all of the time, having no energy or motivation
- Mood changes (feeling sad or angry for no reason)
- Changes in sleep patterns (sleeping a lot more or having a hard time sleeping)
- Light-headedness, dizziness, or loss of balance
- Urge to vomit (nausea)
- Increased sensitivity to lights, sounds, or distractions
- Blurred vision or eyes that tire easily
- Loss of sense of smell or taste
- Ringing in the ears

What are the **two forms** of **stroke**?

There are two forms of stroke: ischemic and hemorrhagic. Ischemic stroke is the blockage of a blood vessel that supplies blood to the brain. Ischemic strokes account for 80 percent of all strokes. Hemorrhagic stroke is bleeding into or around the brain. Hemorrhagic strokes account for 20 percent of all strokes.

What are the **symptoms** of **stroke**?

The symptoms of stroke appear suddenly and include numbness or weakness, especially on one side of the body; confusion or trouble speaking or understanding speech; trouble seeing in one or both eyes; trouble walking, dizziness, or loss of balance or coordination; or severe headache with no known cause. Often more than one of these symptoms will be present, but they all appear suddenly.

How does a **mini-stroke** differ from a regular stroke?

A mini-stroke, technically called a transient ischemic attack (TIA), begins like a stroke but then resolves itself, leaving no noticeable symptoms or deficits. The average duration of a TIA is a few minutes. For almost all TIAs, the symptoms go away within an hour. A person who experiences a TIA should consider it a warning, since approximately one-third of the 50,000 Americans who have a TIA have an acute

> ## How many people in America are estimated to have Alzheimer's disease?
>
> The National Institute on Aging estimates that four million Americans have Alzheimer's disease. Researchers estimate that three percent of people over the aged of 65 have Alzheimer's disease and nearly half of all people over 85 may have the disease.

stroke sometime in the future. Since all stroke symptoms appear suddenly and it is not possible to determine whether it is a TIA or full stroke, medical treatment should be sought immediately.

What are two of the most **common forms** of **dementia**?

The term "dementia" describes a group of symptoms that are caused by changes in brain function. The two most common forms of dementia in older people are Alzheimer's disease and multi-infarct dementia (sometimes called vascular dementia). There is no cure for these types of dementia. In Alzheimer's disease nerve-cell changes in certain parts of the brain result in the death of a large number of cells. Some researchers believe there is a genetic origin to Alzheimer's disease. The symptoms of Alzheimer's disease range from mild forgetfulness to serious impairments in thinking, judgment, and the ability to perform daily activities.

In multi-infarct dementia a series of small strokes or changes in the brain's blood supply may result in the death of brain tissue. The location in the brain where the small strokes occur determines the seriousness of the problem and the symptoms that arise. Symptoms that begin suddenly may be a sign of this kind of dementia. People with multi-infarct dementia are likely to show signs of improvement or remain stable for long periods of time, then quickly develop new symptoms if more strokes occur. In many people with multi-infarct dementia, high blood pressure is to blame.

What are the **seven warning signs** of **Alzheimer's disease**?

The seven warning signs of Alzheimer's disease are:

1. Asking the same question over and over again
2. Repeating the same story, word for word, again and again
3. Forgetting how to cook, how to make repairs, how to play cards, or any other activities that were previously done with ease and regularity
4. Losing one's ability to pay bills or balance one's checkbook
5. Getting lost in familiar surroundings, or misplacing household objects
6. Neglecting to bathe, or wearing the same clothes over and over again, while insisting that one has taken a bath or that clothes are still clean
7. Relying on someone else, such as a spouse, to make decisions

It is important to understand that even if someone has several or even most of these symptoms, it does not mean they definitely have Alzheimer's disease. It does mean they should be thoroughly examined by a medical specialist trained in evaluating memory disorders, such as a neurologist or a psychiatrist, or by a comprehensive memory disorder clinic with an entire team of experts knowledgeable about memory problems.

SPINAL CORD

Where is the **spinal cord located**?

The spinal cord lies inside the vertebral column. It extends from the occipital bone of the skull to the level of the first or second lumbar vertebra. In adults, the spinal cord is 16 to 18 inches (42 to 45 centimeters) long and 0.5 inches (1.27 centimeters) in diameter. The spinal cord is the connecting link between the brain and the rest of the body. It is the site for integration of the spinal cord reflexes.

What is the **cauda equina**?

The cauda equina (from the Latin *cauda,* meaning "tail," and *equus,* meaning "horse") is the group of spinal nerves that arise in the lower lumbar region of the vertebral column. Since the spinal cord extends only to the first or second lumbar vertebrae, all the root nerves that begin below the first or second vertebrae resemble thin, wispy strands of hair similar to a horse's tail.

Why do **physicians perform** a **spinal tap** at the fourth lumbar vertebra (L$_4$)?

A spinal tap, also called a lumbar puncture, is the withdrawal of a small amount of cerebrospinal fluid from the subarchnoid space in the lumbar region of the vertebral column. Since the spinal cord ends at the level of the first or second lumbar vertebra, a needle can be inserted into the subarchnoid space at the fourth lumbar vertebra with little risk of injuring the spinal cord. The cerebrospinal fluid may be tested and examined for infection. Cerebrospinal fluid is also withdrawn to reduce pressure caused by swelling of the brain or spinal cord following injury or disease.

How is the **white matter** in the **spinal cord organized**?

The white matter in the spinal cord is divided into three columns. Each column contains tracts whose axons share functional and structural characteristics.

What are **spinal cord tracts**?

Tracts are bundles of axons that are relatively uniform with respect to diameter, myelination, and conduction speed. All the axons within a tract relay the same type of information in the same direction.

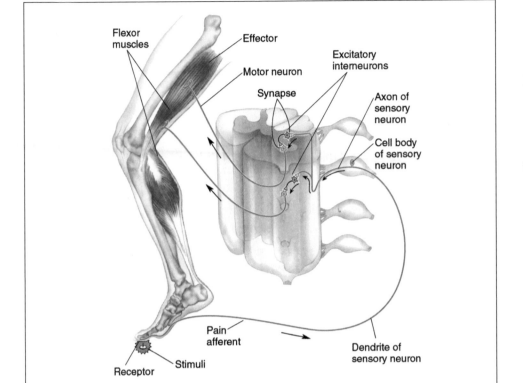

When the body reacts to an outside stimulus without the involvement of the brain, it is called a spinal reflex arc. (From Premkumar K. *The Massage Connection Anatomy and Physiology*. Baltimore: Lippincott, Williams & Wilkins, 2004.)

How do **ascending tracts** differ from **descending tracts**?

Ascending tracts consist of sensory fibers that carry information up the spinal cord to the brain. Descending tracts consist of motor fibers that carry information from the brain to the spinal cord.

What is a **reflex**?

A reflex is a predictable, involuntary response to a stimulus. It was given this name in the 18th century because it appeared that the stimulus was reflected off of the spinal cord to generate the response, just as light is reflected by a mirror. Reflexes allow the body to respond quickly to internal and external changes in the environment in order to maintain homeostasis. Reflexes that involve the skeletal muscles are called somatic reflexes. Reflexes that involve responses of smooth muscle, cardiac muscle, or a gland are called visceral or autonomic reflexes.

What are the components and steps of a **spinal reflex arc**?

Spinal reflex arcs are carried out by neurons in the spinal cord alone, without the immediate involvement of the brain. A reflex arc always starts with a sensory neuron and ends with a motor neuron. The simplest reflex arc is a monosynaptic reflex arc involving two neurons and one synapse. The sensory and motor neurons in a monosynaptic reflex arc synapse directly. Although most reflex arcs are more complex, a monosynaptic reflex arc demonstrates the essential components of a reflex arc.

What are the **steps** in a **knee-jerk reflex arc**?

There are five steps in a knee-jerk reflex arc as follows:

1. A tap on the patellar tendon (tendon attached to the kneecap) is sensed by stretch receptors in the muscle (muscle spindles).

2. The muscle spindle produces a pattern of nerve impulses that are conveyed along a sensory (afferent) nerve fiber, past its cell body in the dorsal root ganglion, and to its termination on a motor neuron.

3. The nerve terminal releases neurotransmitters onto the motor neurons, which generate excitatory, post-synaptic potentials in the motor neurons' dendrites and cell bodies.

4. The motor neuron generates action potentials, leading to the release of acetylcholine from its terminals on muscles.

5. The muscle responds to the acetylcholine by depolarizing and contracting.

How are **reflexes** used to **diagnose diseases** and disorders?

A reflex response other than the normal response may be an indication of a disorder or complication in the nervous system.

Reflex	Description	Disorder or Damage Indicated
Abdominal reflex	Anterior stroking of the sides of lower torso causes contraction of abdominal muscles	Absence of reflex indicates lesions of peripheral nerves or in reflex centers in lower thoracic segments of spinal cord; may also indicate multiple sclerosis
Plantar (Achilles) reflex	Tapping of Achilles tendon of soleus and gastrocnemius muscles causes both muscles to contract, producing plantar flexion of foot	Absence of reflex may indicate damage to nerves innervating posterior leg muscles or to lumbosacral neurons; may also indicate chronic diabetes, alcoholism, syphilis, or subarachnoid hemorrhage
Babinski's reflex	Stroking of lateral part of sole causes toes to curl (plantar reflex)	Reflex indicates damage to upper motor neuron of pyramidal motor system
Biceps reflex	Tapping of biceps tendon in elbow produces contraction of brachialis and biceps muscles, resulting in flexion at elbow	Absence of reflex may indicate damage at the C_5 or C_6 vertebral level

How does the location of a spinal injury on the spinal cord impact the nature of the injuries?

Injuries or damage to the spinal cord at or above the fifth cervical vertebra eliminates sensation and motor control of the upper and lower limbs, as well as any part of the body below the level of the injury. The paralysis after this high spinal injury is termed quadriplegia. Damage that occurs in the thoracic region of the spinal cord effects motor control of the lower limbs only. The paralysis affects only the lower limbs and is called paraplegia.

Reflex	Description	Disorder or Damage Indicated
Brudzinski's reflex	Forceful flexion of neck produces flexion of legs, thighs	Reflex indicates irritation of meninges
Hoffmann's reflex	Flicking of index finger produces flexion in all fingers and thumb	Reflex indicates damage to upper motor neuron of spinal cord
Kernig's reflex	Flexion of hip, with knee straight and patient lying back, produces flexion of knee	Reflex indicates irritation of meninges or herniated intervertebral disc
Patellar reflex (knee jerk)	Tapping of patellar tendon causes contraction of quadriceps fermoris muscle producing upward jerk of leg	Absence of reflex may indicate damage at the L_2, L_3, L_4 vertebral level; may also indicate chronic diabetes, syphilis
Romberg's reflex	Inability to maintain balance when standing with eyes closed	Reflex indicates dorsal column injury
Triceps reflex	Tapping of triceps tendon at elbow causes contraction of triceps muscle, producing extension at elbow	Absence of reflex may indicate damage a C_6, C_7, C_8 vertebral level

How is **trauma** to the **spinal cord classified**?

Injury to the spinal cord produces a period of sensory and motor paralysis termed spinal shock. The severity of the injury will determine how long the paralysis will last and whether there will be permanent damage.

Spinal concussion results in no visible damage to the spinal cord. The resulting spinal shock is temporary and may last for as short as a couple of hours.

Spinal contusion involves the white matter of the spinal cord being injured (bruised). Recovery is more gradual and may involve permanent damage.

125

Spinal laceration, caused by vertebral fragments or other foreign bodies penetrating the spinal cord, often requires a longer recovery and is less complete.

Spinal compression occurs when the spinal cord is squeezed or distorted within the vertebral canal. Relieving the pressure usually relieves the symptoms.

Spinal transection is the complete severing of the spinal cord. Surgical procedures cannot repair a severed spinal cord.

PERIPHERAL NERVOUS SYSTEM: SOMATIC NERVOUS SYSTEM

What are the **components** of the **peripheral nervous system**?

The cranial and spinal nerves constitute the somatic peripheral nervous system. These nerves connect the brain and spinal cord to peripheral structures such as the skin surface and the skeletal muscles.

What is the total **length** of the **peripheral nerves** in the body?

The peripheral nerves measure approximately 93,000 miles (150,000 kilometers) in length.

How many cranial nerves are in the human body?

There are twelve pairs of cranial nerves in the human body. The cranial nerves are designated by Roman numerals and names. The Roman numerals indicate the order in which they emerge from the brain. The name indicates an anatomical feature of function.

What are the **cranial nerves** and their **function**?

Cranial Nerves and Their Function

Name of Cranial Nerve	Function
I Olfactory	Smell
II Optic	Vision
III Oculomotor	Movement of eyeball and eyelid; constricts pupil; focuses lens
IV Trochlear	Movement of eyeball (down and out)
V Trigeminal	Sensations to the face, including scalp, forehead, cheeks, upper lip, palate, tongue, and lower jaw; chewing
VI Abducens	Lateral movement of eye
VII Facial	Facial expressions, taste, secretion of tears, saliva
VIII Vestibulocochlear	Hearing and equilibrium (balance)
IX Glossopharyngeal	Taste and other sensations of the tongue; swallowing and secretion of saliva

Name of Cranial Nerve	Function
X Vagus	Swallowing, coughing, voice production; monitors blood pressure and oxygen and carbon dioxide levels in blood
XI Accessory (also called spinal accessory)	Voice production; skeletal muscles of palate, pharynx, and larynx; movement of head and shoulders
XII Hypoglossal	Movement of tongue during speech and swallowing

Which is the **largest cranial nerve**?

The trigeminal nerve is the largest, although not the longest, cranial nerve.

Which cranial nerve is involved in **tic douloureux**?

Tic douloureux is caused by compression or degeneration of the fifth cranial nerve, the V trigeminal. Individuals afflicted with this condition have sudden, severe, stabbing pain on one side of the face, along the jaw or cheek. The pain may last several seconds and may be experienced repeatedly over several hours, days, weeks, or even months. The episodes may subside as rapidly as they began with no incidents of pain for months or even years.

Which cranial nerve is responsible for **Bell's palsy**?

Bell's palsy, a form of temporary facial paralysis, is the result of damage or trauma to the seventh cranial nerve, the VII facial. The nerve may be swollen, inflamed, or compressed, resulting in an interruption of messages from the brain to the facial muscles. Individuals with Bell's palsy may exhibit twitching, weakness, or paralysis on one or both sides of the face; drooping of the eyelid and corner of the mouth; drooling; dryness of the eye or mouth; impairment of taste; and excessive tearing in one eye. Although the symptoms appear suddenly, individuals begin to recover within two weeks and return to normal function within three to six months.

How many **pairs of spinal nerves** are in the human body?

There are 31 pairs of spinal nerves in the human body. The spinal nerves are grouped according to where they leave the vertebral column. There are eight pairs of cervical nerves (C_1–C_8), twelve pairs of thoracic nerves (T_1–T_{12}), five pairs of lumbar nerves (L_1–L_5), five pairs of sacral nerves (S_1–S_1) and one pair of coccygeal nerves (Co_1).

How are **spinal nerves attached** to the spinal cord?

Spinal nerves divide in the vertebral canal into two branches: the dorsal root and the ventral root. The dorsal root, which is the posterior branch, contains the axons of sensory neurons that bring information to the spinal cord. The ventral root, which is the anterior branch, contains the axons of motor neurons that carry commands to muscles or glands. Therefore, each spinal nerve is considered a mixed nerve with both sensory and motor neurons.

127

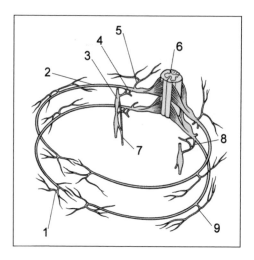

Anatomy of spinal nerves: 1 = Anterior cutaneous branches; 2 = Ventral ramus; 3 = Sympathetic ganglion; 4 = White ramus; 5 = Dorsal ramus; 6 = Spinal cord; 7 = Sympathetic nerve; 8 = Gray ramus; and 9 = Lateral cutaneous branches. (From *Stedman's Medical Dictionary.* 27th Ed. Baltimore: Lippincott, Williams & Wilkins, 2000.)

Which is the **longest spinal nerve**?

The longest spinal nerve is the tibial nerve, which averages 20 inches (50 centimeters) long.

What is a **plexus**?

A plexus (from the Latin *plectere,* meaning "braid") is an interwoven network of spinal nerves. There are four major plexuses on each side of the body: 1) the cervical plexus innervates the muscles of the neck, the skin of the neck, the back of the head, and the diaphragm muscle; 2) the brachial plexus innervates the shoulder and upper limb; 3) the lumbar plexus innervates the muscles and skin of the abdominal wall; and 4) the sacral plexus innervates the buttocks and lower limbs. The nerves then divide into smaller branches.

What are **dermatomes**?

Dermatomes (from the Greek *derma,* meaning "skin," and *tomos,* meaning "cutting") are areas on the skin surface supplied by an individual spinal nerve.

Which **infection** may affect the skin of a single **dermatome**?

Shingles, or herpes zoster, appears as a painful rash on the skin corresponding to the sensory nerve in the area of a single dermatome. The virus is the same one that causes chicken pox. If someone has chicken pox as a child, the virus may lay dormant in the nerve roots of the spinal nerves for decades. If reactivated, the virus will present itself as shingles.

What causes **sciatica**?

Sciatica is caused by compression of the sciatic nerve, such as from a herniated disc or even from sitting for extended periods of time with a wallet in the back pocket. The pain usually subsides after a few weeks, although over-the-counter pain relievers may be helpful.

Which **nerve** is responsible for **carpal tunnel syndrome**?

The median nerve controls sensations to the palm side of the thumb and fingers (although not the little finger), as well as impulses to some small muscles in the hand that allow the fingers and thumb to move. Carpal tunnel syndrome occurs

> ## What causes the sensation of "pins and needles" when your foot "falls asleep"?
>
> Local pressure, such as crossing or sitting on your legs, may temporarily compress a nerve, removing sensory and motor function in your foot. When the local pressure is removed, the familiar feeling of "pins and needles" is felt as the nerve endings become reactivated.

when the median nerve, which runs from the forearm into the hand, becomes pressed or squeezed at the wrist. The carpal tunnel, a narrow, rigid passageway of ligament and bones at the base of the hand, houses the median nerve and tendons. At times thickening from irritated tendons or other swelling narrows the tunnel and causes the median nerve to be compressed. Carpal tunnel syndrome is characterized by pain, weakness, or numbness in the hand and wrist, often radiating up the arm.

According to the National Institute of Neurological Disorders and Stroke, initial treatment of carpal tunnel syndrome includes resting the affected hand and wrist for a minimum of two weeks. Nonsteroidal, anti-inflammatory drugs may be used to ease the pain. Ice and corticosteroids may relieve the swelling and pressure on the nerve. If symptoms persist, surgery may be required to sever the band of tissue around the wrist and reduce pressure on the median nerve.

PERIPHERAL NERVOUS SYSTEM: AUTONOMIC NERVOUS SYSTEM

What does the **autonomic nervous system regulate**?

The autonomic nervous system regulates "involuntary" activity, which is not controlled on a conscious level. Specifically, the autonomic nervous system innervates the activity of smooth muscle, cardiac muscle, and glands of the body.

How is the **autonomic nervous system organized**?

The autonomic nervous system consists of two divisions: the sympathetic nervous system and the parasympathetic nervous system. The sympathetic division is often called the "fight or flight" system because it usually stimulates tissue metabolism, increases alertness, and generally prepares the body to deal with emergencies. The parasympathetic division is considered the "rest and repose" division because it conserves energy and promotes sedentary activities, such as digestion. In general, both the sympathetic and parasympathetic divisions innervate the target cells.

How do the **neural pathways** of the **autonomic nervous system** differ from the **somatic nervous system**?

In the somatic nervous system, the myelinated axon of a motor neuron extends directly from the central nervous system to the effector (e.g., skeletal muscle). Neural pathways in the autonomic nervous system always consist of two neurons. The first neuron, the preganglionic neuron, has its cell body in the central nervous system. Its myelinated axon extends from the central nervous system to an autonomic ganglion, or junction, where it synapses with a second neuron. The second neuron, the postganglionic neuron, is in the peripheral nervous system.

How does the **somatic** nervous system **differ** from the **autonomic** nervous system?

The table below explains the main differences between the somatic and autonomic nervous systems.

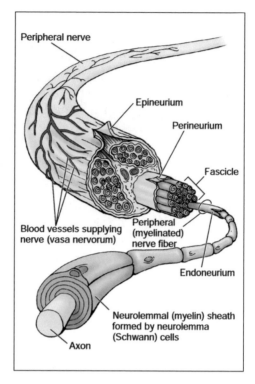

The structure of a peripheral nerve. (From *Stedman's Medical Dictionary*. 27th Ed. Baltimore: Lippincott, Williams & Wilkins, 2000.)

Comparison of Somatic Nervous System and Autonomic Nervous System

	Somatic Nervous System	Autonomic Nervous System
Effectors	Skeletal muscles	Cardiac muscle, smooth muscle, and glands
Type of control	Voluntary	Involuntary
Neural pathway	One motor neuron extends from central nervous system and synapses directly with a skeletal muscle fiber	One motor neuron (preganglion neuron) extends from the central nervous system and synapses with another motor neuron in a ganglion; the second motor neuron (postganglion neuron) synapses with a visceral effector
Neurotransmitter	Acetylcholine	Acetylcholine or norephinephrine
Action of neurotransmitter on effector	Always excitatory (causing contraction of skeletal muscle)	May be excitatory (causing contraction of smooth muscle, increased heart rate, increased force of heart contraction, or increased secretions from glands) or inhibitory (causing relaxation of smooth muscle, decreased heart rate, or decreased secretions from glands)

How does the **effect** of **sympathetic innervation** on an organ differ from the effect of **parasympathetic innervation**?

Many organs are innervated by both the sympathetic and parasympathetic division of the autonomic nervous system.

Comparison of Sympathetic and Parasympathetic Innervation

Effector	Effect of Sympathetic Innervation	Effect of Parasympathetic Innervation
Cardiac muscle		
Heart	Increases heart rate, force of contraction, and blood pressure	Decreases heart rate, force of contraction, and blood pressure
Smooth muscle		
Eye	Dilation of pupil; focusing for distance vision	Constriction of pupil; focusing for near vision
Stomach and intestines	Decreases peristalsis; contracts sphincters	Increases peristalsis; relaxes sphincters
Lungs	Relaxation; airway diameter increases	Contraction; airway diameter decreases
Arrector pili of hair follicles	Contraction that results in erection of hairs ("goose pimples")	No known effect
Urinary bladder	Relaxation of urinary bladder; constricts sphincter	Contraction of muscular wall; relaxation of internal sphincter to eliminate urine
Reproductive organs	Ejaculation of semen in men	Erection of penis (men) or clitoris (women)
Glands		
Sweat	Increases secretion	None (not innervated)
Lacrimal (tears)	None (not innervated)	Stimulates secretion
Salivary	Decreases secretion of digestive juices	Increases secretion of digestive juices
Adrenal	Secretion of epinephrine and norephinephrine by adrenal medullae	None (not innervated)

LEARNING AND MEMORY

Which **parts** of the **brain** are involved in **higher order functions,** such as learning and memory?

Higher order functions, such as learning and memory, involve complex interactions among areas of the cerebral cortex and between the cortex and other areas of the brain. Information is processed both consciously and unconsciously. Since higher order functions are not part of the programmed "wiring" of the brain, the functions are subject to modification and adjustment over time.

What are the **areas** of the **cerebral cortex** and their functions?

The cerebral cortex is divided into three functional areas: 1) sensory areas, 2) motor areas, and 3) association areas. The sensory areas receive and interpret sensory impulses. The motor areas control muscular movement. The association areas are involved in integrative functions such as memory, emotions, reasoning, will, judgment, personality traits, and intelligence.

Which **areas** of the **brain** are responsible for **specific functions**?

Researchers know that certain areas of the brain are responsible for certain general functions. In 1909, the German physician and researcher Korbinian Brodmann (1868–1918), published *Vergleichende Lokalisationslehre der Grosshirnrinde in ihren Prinzipien dargestellt auf Grund des Zellenbaues*. This treatise included maps of the localization of functions in the cerebral cortex. Brodmann's maps are still used to depict the areas of cerebral cortex that are responsible for specific functions.

Some Functionally Important Brodmann's Areas

Area	Function
1, 2, 3	Primary somatosensory area (touch, joint and muscle position, pain, temperature)
4	Primary motor area (controls specific muscles or groups of muscles)
5, 7	Somatosensory association area (integrates and interprets somatic sensations; also stores memories of past sensory experiences)
6	Premotor area (deals with learned motor activities of a complex and sequential nature)
8	Frontal eye field area (eye movements)
9, 10, 11	Tertiary motor movement
17, 18, 19, 20, 21	Vision (conveys visual information about shape and color; interpreting and evaluating visual experiences)
22	Auditory association area (interprets sound as speech, music, or noise)
28	Primary olfactory area (receives impulses for smell)
39, 40 (also 22)	Wernicke's area (speech)
41, 42	Primary auditory area (receives impulses for hearing the characteristics of sound, such as pitch and rhythm)
43	Primary gustatory area (receives impulses for taste)
44, 45	Broca's area (speech)

Who discovered which **area** of the **brain** is responsible for **speech and language**?

Pierre Paul Broca (1824–1880) identified an area of the brain responsible for speech production in 1861. Having observed a patient who could not speak except for the meaningless utterance of "tan, tan," Broca examined the brain of the individual upon his death. Broca determined the patient was missing a section of the frontal lobe from the cerebral hemisphere. Broca continued to examine the brains of individuals with a lack of speech and found they all lacked the same area of the brain.

There is still an ongoing debate about how to measure intelligence and whether inherited traits or the environment affect learning abilities more. (© iStockphoto.com/LisaFX Photographic Designs.)

How do **Broca's area** and **Wernicke's area** differ?

Broca's area and Wernicke's area are both associated with speech. Broca's area is associated with the production of speech. It controls the flow of words from brain to mouth. Wernicke's area is associated with the interpretation and understanding of speech.

What are the causes of **aphasia**?

Aphasia is a language disorder that results from damage to portions of the brain that are responsible for language. Strokes are the most common cause of aphasia, although aphasia can also result from a brain tumor, infection, head injury, or dementia that damages the brain. Individuals with aphasia have difficulty speaking—both in producing words and complete sentence structure—or understanding speech, or both.

Depending on the severity of the aphasia (and the degree of permanent brain damage), some patients regain their speech capabilities with little or no rehabilitation. In most cases, however, speech therapy is necessary to regain language capabilities.

Is it possible to **measure intelligence**?

The earliest test created to measure intelligence was developed by French physiologist Alfred Binet (1857–1911) in 1905. The purpose of the test was to measure skills such as judgment, comprehension, and reasoning in order to place children in the appropriate classes in school. The test was brought to the United States by Stanford University psychologist Lewis Terman (1877–1956) in 1916 and renamed the Stanford-Binet test. Since then, other intelligence tests, such as the Wechsler Adult Intelligence Scale and the Wechsler Intelligence Scale for Children have been developed. These tests have produced a score referred to as an intelligence quotient or IQ.

How is **IQ calculated**?

IQ, or the intelligence quotient, was originally computed as the ratio of a person's mental age to his chronological age, multiplied by 100. Following this method, a child of 10 years old who performed on the test at the level of an average 12 year old (mental age of 12), was assigned an IQ of $12/10 \times 100 = 120$. More recently, the concept of "mental age" has fallen into disrepute and IQ is computed on the basis of the statistical percentage of people who are expected to have a certain IQ. An IQ of 100 is considered average. An IQ of 70 or below indicates mental retardation, and an IQ of 130 or above indicates gifted abilities.

What is **memory**?

Memory is the ability to recall information and experiences. Memory and learning are related because in order to be able to remember something it must first be "learned." Memories may be facts or skills. Memory "traces" have been described traditionally as concrete things that are formed during learning and imprinted on the brain when neurons record and store information. However, the way that memories are formed and represented in the brain is not well understood.

How does **short-term memory** differ from **long-term memory**?

Short-term memory, also called primary memory, refers to small bits of information that can be recalled immediately. The recalled information has no permanent importance, such as a name or telephone number that is only used once. Long-term memory is the process by which information that for some reason is interpreted as being important is remembered for a much longer period. Short-term memories may be converted to long-term memories.

Which **areas** of the **brain** are involved in **memory**?

Several areas of the brain are associated with memory, including the association cortex of the frontal, parietal, occipital, and temporal lobes, the hippocampus, and the diencephalon. Damage to the hippocampus results in an inability to convert short-term memories to long-term memories. Memory loss may be the result of trauma or injury, disease, lifestyle choices, such as alcoholism and drug use, and aging.

What is **amnesia**?

Amnesia refers to the loss of memory from disease or trauma. The extent and type of memory loss is dependent on the area of the brain that is damaged. Individuals with retrograde amnesia suffer from memory losses of past events. This is a common occurrence when head injury is involved. Oftentimes, an individual will not be able to recall the events and moments immediately preceding an accident or fall.

Individuals who suffer from anterograde amnesia are unable to store additional memories, but their earlier memories are intact and accessible. They have difficulty creating new long-term memories. As a consequence, every experience is a new experience for these individuals, even if they have experienced it earlier, such as meeting a new person or reading a new book.

SLEEP AND DREAMS

What is **consciousness**?

A conscious individual is alert and attentive to his or her surroundings, while an unconscious individual is not aware of his or her surroundings. Conscious states, however, range from normal consciousness to the conscious yet unresponsive state, while unconscious states range from being asleep to being in a coma.

How does the **Glasgow coma scale** classify the level of consciousness?

The Glasgow coma scale is the most widely used system of classifying the severity of head injuries or other neurologic diseases. It rates three areas of response, involving eye, verbal, and motor responses, and then tallies a total score. The following chart details the scale:

When was the term "brain death" first coined?

The clinical state known as "brain death" was first described as a *coma dépassé* (literally, a state beyond coma) by French neurologists P. Mollaret and M. Goulon in 1958.

Glasgow Coma Scale

Area of Assessment	Response	Score
Eye opening response	Spontaneous—open with blinking at baseline	4
	To verbal stimuli, command, speech	3
	To pain only (not applied to face)	2
	No response	1
Verbal response	Oriented and able to converse	5
	Confused conversation, but able to answer questions	4
	Inappropriate words	3
	Incomprehensible speech	2
	No response	1
Motor response	Obeys commands for movement	6
	Purposeful movement to painful stimulus	5
	Withdraws in response to pain	4
	Flexion in response to pain (decorticate posturing)	3
	Extension response in response to pain (decerebrate posturing)	2
	No response	1

Total Glasgow coma scores (GCS) indicate the level of the coma:

- GCS of 3 to 8: a coma, categorized as no eye opening, no ability to follow commands, no word verbalizations
- GCS of 8 or less: severe head injury
- GCS of 9 to 12: moderate head injury
- GCS of 13 to 15: mild head injury

What are the **stages of sleep**?

Data collected from EEGs (electroencephalograms) of brain activity during sleep have shown at least four separate stages of sleep. During stage 1, heart and breathing rates decrease slightly, the eyes roll slowly from side to side, and an individual experiences a floating sensation. Stage 1 sleep is not usually classified as "true" sleep. This stage generally lasts only five minutes. Individuals awakened during stage 1 sleep will often insist that they were not sleeping, but merely "resting their eyes."

Stage 2 sleep is characterized by the appearance of short bursts of waves known as "sleep spindles" along with "K complexes," which are high-voltage bursts that

Lack of sufficient sleep has a definite effect on a person's ability to function during the day. © iStockphoto.com/Digital Savant LLC.

occur before and after a sleep spindle. Eyes are generally still and heart and breathing rates decrease only slightly. Sleep is not deep.

Stage 3 sleep is intermediate sleep and is characterized by steady, slow breathing, a slow pulse rate, and a decline in temperature and blood pressure. Only a loud noise awakens sleepers in stage 3 sleep.

Stage 4 sleep, known as oblivious sleep, is the deepest stage. It usually does not begin until about an hour after falling asleep. Brain waves become even slower, and heart and breathing rates drop to 20 or 30 percent below those in the waking state. The sleeping individual in stage 4 sleep is not awakened by external stimuli, such as noise, although an EEG will indicate that the brain acknowledges such stimuli. Stage 4 sleep continues for close to an hour, after which the sleeper will gradually drift back into stage 3 sleep, followed by stages 2 and then 1, before the cycle begins again.

Why do people **need sleep**?

Scientists do not know exactly why people need sleep, but studies show that sleep is necessary for survival. Sleep appears to be necessary for the nervous system to work properly. While too little sleep one night may leave us feeling drowsy and unable to concentrate the next day, a prolonged period of too little sleep leads to impaired memory and physical performance. Hallucinations and mood swings may develop if sleep deprivation continues.

137

What is **REM** sleep?

REM sleep is rapid eye movement sleep. It is characterized by faster breathing and heart rates than NREM (nonrapid eye movement) sleep. The only people who do not have REM sleep are those who have been blind from birth. REM sleep usually occurs in four to five periods, varying from five minutes to about an hour, growing progressively longer as sleep continues.

What is the **sleep cycle**?

Typically, there are several cycles of sleep each night. Each cycle begins with a period of REM sleep. Earlier in the night there will be periods of stage 3 and stage 4 sleep, but these diminish towards morning, when there are longer periods of REM sleep and less deep sleep.

When does **dreaming occur** during the sleep cycle?

Almost all dreams occur during REM sleep. Scientists do not understand why dreaming is important, one theory is that the brain is either cataloging the information it acquired during the day and discarding the data it does not want, or is creating scenarios to work through situations causing emotional distress. Regardless of its function, most people who are deprived of sleep or dreams become disoriented, unable to concentrate, and may even have hallucinations.

Why is it difficult to **remember dreams**?

It appears that the content of dreams is stored in short-term memory and cannot be transferred into long-term memory unless they are somehow articulated. Sleep studies show that when individuals who believe they never dream are awakened at various intervals during the night when they are in the middle of a dream.

How much sleep does an individual **need**?

As a person ages, the time spent sleeping changes. The following table shows how much sleep is generally needed at night, depending on age.

Age	Sleep Time (in hours)
1–15 days	16–22
6–23 months	13
3–9 years	11

Age	Sleep Time (in hours)
10–13 years	10
14–18 years	9
19–30 years	8
31–45 years	7.5
45–50 years	6
50 or more years	5.5

By the time someone is 20 years old they will have spent approximately eight years of their life asleep; by age 60, they will have spent about twenty years sleeping.

How long can a person **survive without sleep**?

A total lack of sleep will cause death quicker than starvation. A person can survive a few weeks without food, but only ten days without sleep. Sleep deprived individuals experience extreme psychological discomfort after a few days, followed by hallucinations and psychotic behavior.

What are some **sleep disorders**?

The most common sleep disorder is insomnia. Insomnia is ongoing difficulty in falling asleep, staying asleep, or restless sleep. Technically, insomnia is a symptom of other sleep disorders. Consequently, treatment for insomnia depends on the primary cause of insomnia, which may be stress, depression, or too much caffeine or alcohol.

Hypersomnia is extreme sleepiness during the day even with adequate sleep the night before. Hypersomnia has been mistakenly blamed on depression, laziness, boredom, or other negative personality traits.

Narcolepsy is characterized by falling asleep at inappropriate times. The sleep may last only a few minutes and is often preceded by a period of muscular weakness. Emotional events may trigger an episode of narcolepsy. Some individuals with narcolepsy experience a state called sleep paralysis. They wake up to find their body is paralyzed except for breathing and eye movement. In other words, the brain is awake but the body is still asleep.

Sleep apnea is a breathing disorder in which an individual briefly wakes up because breathing has been interrupted and may even stop for a brief period of time. Obstructive sleep apnea (OSA) is the most common form of sleep apnea. It occurs when air cannot flow into or out of the person's nose or mouth as they breathe.

When does **sleepwalking occur** during the sleep cycle?

Sleepwalking generally occurs during deep sleep, but may also be present during periods of NREM (non-rapid eye movement) sleep. It is most common in children, although the National Sleep Foundation estimates that 1 percent to 15 percent of the population may be sleepwalkers. Sleepwalkers generally remain asleep and do not remember leaving their beds. Contrary to popular myth, sleepwalkers should be awakened, although they may be confused when awakened.

What are **circadian rhythms**?

Circadian (from the Latin *circa,* meaning "about," and *dies,* meaning "day") are the regular, internal body rhythms. Although our lives revolve around a 24-hour day, researchers have found that normal circadian rhythms are more on a 25-hour cycle. Many physiological processes, including the sleep/wake cycle, body temperature, gastric secretion, and kidney function, follow a set pattern. For example, body temperature peaks in the late afternoon/early evening and is lowest between 2:00 A.M. and 5:00 A.M. Blood pressure, heartbeat, and respiration follow rhythmical cycles. The production of urine drops at night, allowing for uninterrupted rest.

Circadian rhythm disturbances occur when sleep/wake cycles are interrupted. They often affect shift workers whose biological clocks are disrupted by conflicting sleep and work schedules. "Jet lag" is another form of circadian rhythm disturbance.

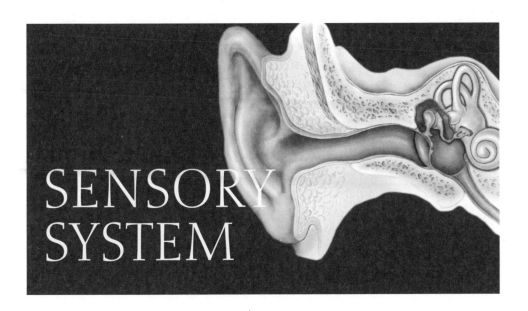

SENSORY SYSTEM

INTRODUCTION

What are the **major senses**?

As early as the days of the Greek philosopher Aristotle, the five senses were recognized to include smell, taste, sight, hearing, and touch. More recently, scientists categorize the senses into two major groups. One group is the special senses, which are produced by highly localized sensory organs and include the senses of smell, taste, sight, hearing, and balance. The other group is the general senses, which are more widely distributed throughout the body and include such senses as touch, pressure, pain, temperature, and vibration.

What are **sensory receptors**?

Sensory receptors are structures in the skin and other tissues that detect changes in the internal or external environment. These receptors consist of specialized neuron endings or specialized cells in close contact with neurons that convert the energy of the stimulus (sound, color, odor, etc.) to electrical signals within the nervous system. Sensory receptors, together with other cells, compose the major sense organs, including eyes, ears, nose, and taste buds.

How many **types** of **sensory receptors** have been identified?

Five types of sensory receptors, each responding to a different type of stimulus, have been identified.

Chemoreceptors—Respond to chemical compounds such as odor molecules.

Photoreceptors—Respond to light.

Thermoreceptors—Respond to changes in temperature.

Mechanoreceptors—Respond to changes in pressure or movement.

Pain receptors—Respond to stimuli that result in the sensation of pain.

Are **sensory perceptions** the same in **men and women**?

Studies have shown that women are more sensitive to smell, taste, touch, hearing, and vision than men and that they are influenced more clearly by hormonal factors. Measuring the acuity of a woman's senses during a monthly cycle found them most acute at ovulation, when estrogen was at its highest level.

What **structures in the body** are associated with the receptors for the **general senses**?

Receptors for the general senses, including touch, pressure, pain, temperature, and vibration, are usually associated with the skin. However, others are associated with deeper structures, such as tendons, ligaments, joints, muscles, and viscera.

Is the **tongue** a **sensitive organ**?

The tongue is more sensitive to touch, temperature, and pain than any other part of the body.

What is **another name** for **pain receptors**?

Pain receptors are also known as nociceptors. They are especially common in the superficial portions of the skin, in joint capsules, within the periosteum of bones, and around the walls of blood vessels.

What **sense** is most closely associated with **emotions**?

Smell is the sense that is most closely tied to our emotions. Because some of the nerves that travel to the brain from olfactory receptors must pass through the limbic system, thus stimulating it and its centers of emotion and sexuality each time a smell is received.

What is phantom limb pain?

Phantom limb pain is perceived in tissue that is no longer present. This name was attached to the phenomenon by a physician during the American Civil War, when a veteran with amputated legs asked for someone to massage his cramped leg muscle. One explanation for phantom limb pain is that the nerves remaining in the stump may generate nerve impulses that are transmitted to the brain and interpreted as arising from the missing limb. Other theories propose that the phantom sensation might be produced by brain reorganization caused by the absence of the sensations that would normally arise from the missing limb.

How are **variations in temperature detected** by the body?

Temperature sensations are detected by specialized free nerve endings called cold receptors and warm receptors. Cold receptors respond to decreasing temperature and warm receptors respond to increasing temperatures. Cold receptors are most sensitive to temperatures between 50°F (10°C) and 68°F (20°C). Temperatures below 50°F (10°C) stimulate pain receptors, producing a freezing sensation. Warm receptors are most sensitive to temperatures above 77°F (25°C) and become unresponsive at temperatures above 113°F (45°C). Temperatures near and above 113°F (45°C) stimulate pain receptors, producing a burning sensation. Both warm and cold receptors rapidly adapt. Within about a minute of continuous stimulation, the sensation of warmth or cold begins to fade.

SMELL

How does the **sense of smell work**?

The sense of smell is associated with sensory receptor cells in the upper nasal cavity. The smell, or olfactory, receptors are chemoreceptors. Chemicals that stimulate olfactory receptors enter the nasal cavity as airborne molecules called gases. They must dissolve in the watery fluids that surround the cilia of the olfactory receptor cells before they can be detected. These specialized cells, the olfactory receptor neurons, are the only parts of the nervous system that are in direct contact with the outside environment. The odorous gases then waft up to the olfactory cells, where the chemicals bind to the cilia that line the nasal cavity. That action initiates a nerve impulse being sent through the olfactory cell, into the olfactory nerve fiber, to the olfactory bulb, and to the brain. The brain then knows what the chemical odors are.

How does the **brain detect different smells**?

The exact mechanism of olfaction is unclear, but some studies have been completed that attempt to classify smells into groups such as floral, mushy, pungent, etc.,

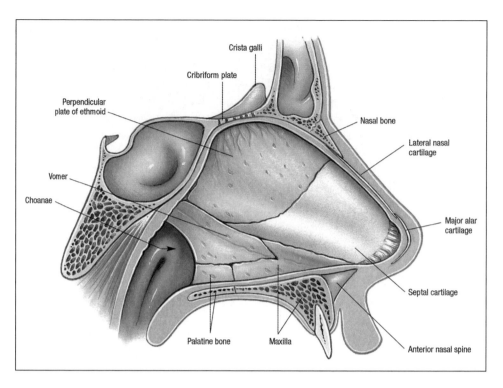

The parts of the nose. *Anatomical Chart Co.*

which are related to molecules of a particular shape and size. One hypothesis suggests that the shapes of gaseous molecules fit complementary shapes of membrane receptor sites on those olfactory receptor cells.

How does **aging affect** the sense of **smell**?

Since the olfactory receptor neurons are exposed to the external environment, they are subject to damage over time. People typically experience a progressive diminishing of olfactory sense with age. In fact, it is estimated that an individual loses about one percent of the olfactory receptors every year.

Can **olfactory cells** be **replaced**?

Yes, injured olfactory cells may be replaced. Small basal cells, located in the olfactory epithelial layer, are capable of dividing and differentiating into olfactory receptor cells. These cells act as a group of neural stem cells.

Why is it possible to **smell** (or "**taste**") medications, such as **eye drops** that have been placed in the eyes?

Medications placed in the eyes can pass through the nasolacrimal duct into the nasal cavity, where their odor can be detected. Because much of our taste sensation is actually smell, the medication is perceived to have a taste.

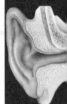

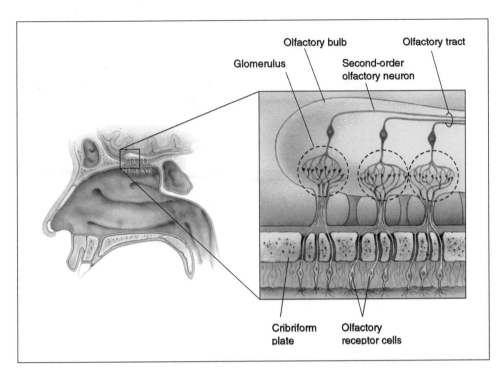

Olfactory nerves in the nose are what allow people to sense smells. (From Bear, M.F., Connors, B.W., and Parasido, M.A. *Neuroscience: Exploring the Brain*. 2nd Ed. Philadelphia: Lippincott, Williams & Wilkins, 2001.)

Do **humans or bloodhounds** have a **keener** sense of **smell**?

Humans smell the world using about 12 million olfactory receptor cells, whereas bloodhounds have 4 billion such cells and, therefore, a much better sense of smell. For example, the trace of sweat that seeps through shoes and is left in footprints is a million times more powerful than the bloodhound needs to track down someone.

What is **anosmia**?

Anosmia is a partial or complete loss of smell, either on a temporary basis or permanently. It may result from a variety of factors, including inflammation of the nasal cavity lining due to a respiratory infection, excessive tobacco smoking, or from the use of certain drugs such as cocaine. In young people, the loss of smell is most often linked to a viral infection, while in the elderly, it more commonly follows a head injury.

Which **diseases** can be **detected by smell**?

Many diseases give off a characteristic odor all their own. Some physicians can detect various diseases by smelling their patients.

Condition or Disease	Odor
Arsenic poisoning	Garlic
Some types of cancers	Fetid

Condition or Disease	Odor
Coma and diabetes	Sweat (acetone)
Coma and kidney malfunction	Ammonia-like
Coma and bowel obstruction	Feces
Diptheria	Sickeningly sweet
Eczema and impetigo	Moldy
Measles	Freshly plucked feathers
Plague	Apples
Pseudomonas infection	Musty wine cellar
Scurvy	Putrid
Smallpox	Putrid
Typhoid fever	Fresh-baked bread
Yellow fever	Butcher shop

TASTE

What are the **special organs** of **taste**?

The special organs of taste are the taste buds located primarily on the surface of the tongue, where they are associated with tiny elevations called papillae surrounded by deep folds. A taste bud is a cluster of approximately 100 taste cells representing all taste sensations and 100 supporting cells that separate the taste cells. Taste buds can also be found on the roof of the mouth and in the throat. An adult has approximately 10,000 taste buds.

What is the **average lifespan** of a human **taste bud**?

Each taste bud lives for seven to ten days.

How do **taste buds function**?

The taste cells that comprise each taste bud act as receptors. Taste cells and adjacent epithelial cells comprise a spherical structure with small projections called taste hairs that protrude from the taste cells. The taste hairs are the sensitive part of each receptor cell. A network of nerve fibers surrounds and connects all of the taste cells. Stimulation of a receptor cell triggers an impulse on a nearby nerve fiber, and the impulse then travels to the brain via a cranial nerve for interpretation.

How many **basic taste sensations** are recognized?

It has been believed generally that there are only four basic taste sensations: sweet, sour, salty, and bitter. Some other taste sensations that are frequently mentioned are alkaline, metallic, and umami, which detects monosodium glutamate (MSG), a flavor enhancer often used in Chinese cooking. Different tastes are experienced by combining the four basic taste sensations. Some individuals claim that with the

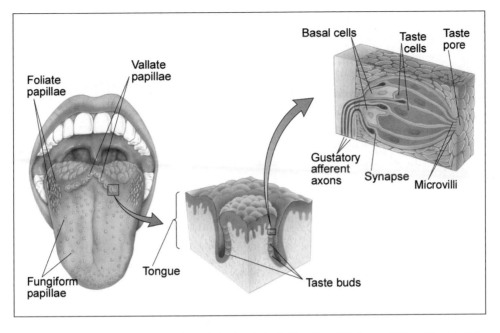

The tongue is covered in taste buds that send neural signals to the brain. (From Bear, M.F., Connors, B.W., and Parasido, M.A. *Neuroscience: Exploring the Brain*. 2nd Ed. Philadelphia: Lippincott, Williams & Wilkins, 2001.)

senses of smell and taste working together, an individual can experience 10,000 different combinations.

What is **umami**?

Umami, a new taste described by the Japanese, is elicited by the amino acid glutamate, which appears to be responsible for the "beef taste" of steak, the tang of aging cheese, and the flavor of MSG, the food additive monosodium glutamate. The umami receptors are located in the pharynx.

Are certain **areas of the tongue** associated with a **particular taste** sensation?

All taste buds are able to detect each of the four basic taste sensations. However, each taste bud is usually most sensitive to one type of taste stimuli. The stimulus type to which each taste bud responds most strongly is related to its position on the tongue. Sweet receptors are concentrated at the tip of the tongue, while sour receptors are more common at the sides of the tongue. Salt receptors occur most frequently at the tip and front edges of the tongue. Bitter receptors are most numerous at the back of the tongue.

Does the sense of **taste diminish with age** like the sense of smell?

The sense of taste diminishes with age, but not as significantly as the sense of smell. The number of taste buds begins to decrease around age 50 and, therefore, taste and flavor perception declines. According to one study, by age 60, most people have lost

The special senses of smell and taste are very closely related, both structurally and functionally. Experimental evidence shows that taste is partially dependent on the sense of smell. Most subjects are unable to distinguish
between an onion and an apple on a blind taste test when their sense of smell
is blocked. This also explains why food is "tasteless" when you have a cold
because the olfactory receptor cells are covered with a thick mucus blocking
the sense of smell.

half of their taste buds. Perhaps this is why older people season their food rather
heavily. Older people often lose their ability to sense bitter or salty flavors entirely.

HEARING

What **two functions** are performed by the **ear**?

The ear has two functions: hearing and maintaining equilibrium or balance.

What are the **major parts** of the **ear**?

The major parts of the ear are external ear, middle ear, and inner ear.

What are the **parts** of the **external ear**?

The external ear is the visible part of the ear. It consists of an outer, funnel-shaped structure called the auricle (pinna) and a tube called the auditory canal that leads inward for
about 1 inch (2.5 centimeters). It ends at the eardrum (tympanic membrane).

How **thick** is the **eardrum**?

The eardrum is 0.00435 inches (about 0.11 millimeters) thick.

What **structures** comprise the **middle ear**?

The middle ear consists of the tympanic membrane (eardrum), tympanic cavity (an
air-filled space in the temporal bone), and three small bones called auditory ossicles.
The tympanic cavity is connected to the nasopharynx (the region linking the back of
the nasal cavity and the back of the oral cavity) by the auditory (Eustachian) tube.

What are the **three bones** in the **middle ear**?

The three bones, or auditory ossicles, in the middle ear are the malleus (hammer),
the incus (anvil), and stapes (stirrup). Tiny ligaments attach them to the wall of the

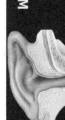

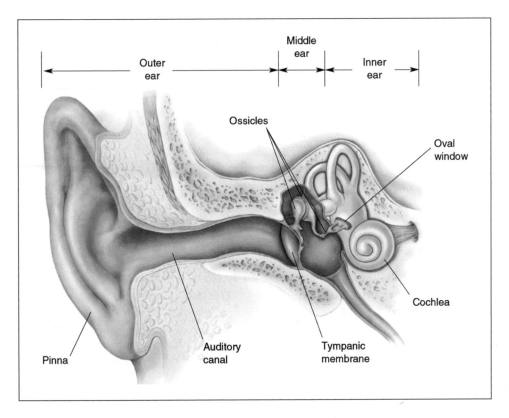

The parts of the ear. (From Bear, M. F., Connors, B. W., and Parasido, M. A. *Neuroscience: Exploring the Brain*. 2nd Ed. Philadelphia: Lippincott, Williams & Wilkins, 2001.)

tympanic cavity, and they are covered by mucous membranes. A special muscle, the stapedius, is attached to the stapes and can dampen its vibrations. These bones bridge the eardrum and the inner ear, transmitting vibrations.

What is the **auditory tube** and its function?

The auditory tube (Eustachian tube) connects each middle ear to the throat. This tube conducts air between the tympanic cavity and the outside of the body by way of the throat and mouth. It also helps maintain equal air pressure on both sides of the eardrum, which is necessary for normal hearing. The function of the auditory tube can be experienced during rapid change in altitude. As a person moves from a high altitude to a lower one, the air pressure on the outside of the membrane becomes greater and greater. As a result, the eardrum may be pushed inward, out of its normal position, and hearing may be impaired. When the air pressure difference is great enough, some air may force its way up through the auditory tube into the middle ear. This allows the pressure on both sides of the eardrum to equalize, and the drum moves back to its regular position. An individual usually hears a popping sound at this time, and normal hearing is restored. A reverse movement of air occurs when a person moves from a low altitude to a higher one.

149

What is the **labyrinth**?

The labyrinth is a complex system of chambers and tubes in the inner ear. There are actually two labyrinths in each ear: the osseous—or bony—labyrinth and the membranous labyrinth. The three regions of the bony labyrinth are the vestibule, the cochlea, and the semicircular canals. There are two membranous sacs within the vestibule, the saccule and utricle, which contain receptors that respond to linear acceleration (e.g., the pull of gravity, acceleration in a vehicle, and changes in head position).

What are the **basic stages** of **sensing sound**?

A sound wave is a vibration in the air that enters the ear canal. The sound strikes the eardrum, causing it to vibrate. Behind the vibrating eardrum, in the middle ear, are three small bones that move in response to the eardrum. These bones transfer the vibrations to the cochlea, traveling through the cochlear duct toward the auditory nerve. Nerve impulses travel to the brain, which translates them into a sound you can understand.

What is the **organ of Corti**?

The organ of Corti, located in the cochlear duct, is the auditory organ. It contains about 20,000 hearing receptor cells and many supporting cells. These receptor cells are called hair cells. The organ of Corti sits on the basilar membrane, a flexible, fibrous structure on the floor of the cochlear duct. As a pressure wave travels through the cochlear duct it causes the basilar membrane to vibrate. The basilar membrane is narrow and stiff at the base of the cochlea (like the strings of a harp or piano used in playing the high notes), where it resonates in response to high-frequency sound waves. The basilar membrane is wide and less stiff near the apex of the cochlea (like the strings of a harp or piano for the low notes), where it resonates in response to lower-frequency pressure waves. This resulting vibration causes the organ of Corti to vibrate, which is sensed by hair cells. Depending on the volume of the sound, either a few hairs move, as in the case of a soft sound, or many hairs move, as in the case of the loud sound.

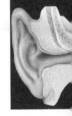

What **word** is used to describe **sounds too low for humans** to hear?

Sound waves are alternating zones of high and low pressure traveling through air or water and characterized by their frequency or intensity. Frequency is measured in hertz (Hz), which represents cycles per second (cps). The frequency range of human hearing is from 20 to 20,000 Hz. Sounds with a frequency lower than 20 Hz cannot be heard by humans and are referred to as infrasonic. Such signals start below 20 Hz, but can be detected at frequencies as low as a hundredth or even a thousandth of a hertz. Human ears are most sensitive to frequencies between 1,500 and 4,000 Hz. Within that range, an individual can distinguish frequencies that differ by only 2 or 3 Hz.

What are some common **levels of sound** and how do they **affect hearing**?

Some common levels of sound and their effects on hearing are listed in the following table:

Sound	Decibel Level	Affects on Hearing
Lowest audible sound	0	None
Rustling leaves	20	None
Quiet library or office	30–40	None
Normal conversation; refrigerator running; light road traffic at a distance	50–60	None
Busy car traffic; vacuum cleaner; noisy restaurant	70	None
Heavy city traffic; subway; shop tools; power lawn mower	80–90	Some damage if continuous for 8 hours or more
Chain saw	100	Some damage if continuous for 2 hours
Rock concert	110–120	Definite risk of permanent hearing loss
Gunshot	140	Immediate danger of hearing loss
Jet engine	150	Immediate danger of hearing loss
Rocket launching pad	160	Hearing loss inevitable

What are the **two types** of **deafness**?

The two types of deafness are conduction deafness and sensorineural, or perceptive, deafness. In conduction deafness, the transmission of sound waves through the middle ear is impaired. In sensorineural deafness, the transmission of nerve impulses anywhere from the cochlea to the auditory cortex of the brain is impaired.

What are some **causes** of **hearing loss** and **deafness**?

Deafness may be caused by dysfunction of either the sound-transmitting mechanism of the outer, middle, or inner ear, or the sound-receiving mechanism of the

151

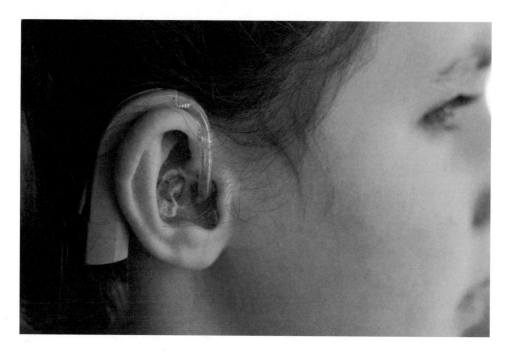

Most people associate hearing loss with age, but there are many causes that can affect even young children. © iStockphoto.com/fotoIE.

inner ear. Causes of dysfunction include disease, toxic exposure, injury (including exposure to loud noise such as heavily amplified music through headphones), or genetic disorders.

What is **presbycusis**?

Presbycusis is the scientific name for age-related sensorineural hearing loss. The first symptom is an inability to hear sounds at the highest frequencies and can occur as early as age 20. Around age 60, there is considerable variation in how well people hear. Some have had significant loss of hearing since age 50, while others have no hearing problems into their 90s. In general, men seem to experience hearing loss more often and more severely than women. One explanation for this difference may be that men's occupations are usually associated with prolonged exposure to louder noises.

What is **tinnitus**?

Tinnitus is the perception of sound in the ears or head where no external source is present. In almost all cases, tinnitus is a subjective noise, meaning that only the person who has tinnitus can hear it. It is often referred to as "ringing in the ears." Persistent tinnitus usually indicates the presence of hearing loss. The exact cause of tinnitus is not known, but there are several likely sources, all of which are known to trigger or worsen the condition. They include noise-induced hearing loss, wax build-up in the ear canal, medicines that are toxic to the ear, ear or sinus infections, jaw misalignment, and head and neck trauma.

Where are the **organs of equilibrium** located?

The organs of equilibrium are located in the inner ear. The otolith organs are located in the vestibule of the membranous labyrinth. They consist of sheets of hair cells covered by a membrane that contains otoliths ("ear stones"), which are calcium carbonate crystals. The otolith organs sense linear acceleration of the head in any direction, such as acceleration due to changing the position of your head relative to gravity or acceleration in a car or amusement ride. The inner ear also contains horizontal, posterior, and anterior semicircular canals, which sense angular motions (acceleration) of the head. Each semicircular canal has a specialized sensory region that contains hair cells, and each canal is important for sensing rotation of the head in a different primary direction. For example, the horizontal semicircular canal receptors are sensitive to rotating the head leftward and rightward.

What is **Meniere's disease**?

Meniere's disease, named after Prosper Meniere (1799–1862), who first described it in 1861, is a disorder characterized by recurring attacks of disabling vertigo (a whirling sensation), hearing loss, and tinnitus. It is thought to be caused by an imbalance in the fluid that is normally present in the inner ear. Either an increase in the production of inner ear fluid or a decrease in its reabsorption results in an imbalance of fluid, but why this happens is not known. It most often occurs in middle age and is more common in men than women.

What is **motion sickness**?

Motion sickness (also known as car, sea, train, or air sickness) occurs when the body is subjected to accelerations of movement in different directions or under conditions where visual contact with the actual outside horizon is lost. The brain receives contradictory information from its motion sensors such as the eyes or semicircular canals in the middle ears that provide information about body position. Symptoms include dizziness, fatigue, and nausea, which may progress to vomiting. Prevention is best accomplished by seeking areas of lesser movement in an interior location of a ship or by facing forward and looking outside an airplane. Various prescription and over-the-counter medications are available that may prevent or limit the symptoms of motion sickness.

What is **cerumen**?

Cerumen is an oily, fatty substance produced by the ceruminous glands in the outer portion of the ear canal. This compound is commonly referred to as ear wax and, together with hairs in the auditory canal, helps prevent foreign objects from reaching the delicate eardrum. Dust, dirt, bacteria, fungi, and other foreign dangers to the body all stick to the wax and do not enter the ear. Ear wax also contains a special enzyme, lysozyme, which breaks down the cell walls of bacteria.

Should **ear wax** be **removed**?

In most individuals, the ear canal is self-cleansing and there is no need to remove ear wax. However, ear wax may be impacted due to poor attempts at cleaning the ear. In such cases, the impacted ear wax should be removed by a healthcare professional.

VISION

What are the **parts of the eye** and their **functions**?

The major parts of the eye and their functions are summarized in the following chart:

Structure	Function
Sclera	Maintains shape of eye; protects eyeball; site of eye muscle attachment
Cornea	Refracts incoming light; focuses light on the retina
Pupil	Admits light
Iris	Regulates amount of incoming light
Lens	Refracts and focuses light rays
Aqueous humor	Helps maintain shape of eye; maintains introcular pressure; nourishes and cushions cornea and lens
Ciliary body	Holds lens in place; changes shape of lens
Vitreous humor	Maintains intraocular pressure; transmits light to retina; keeps retina firmly pressed against choroids
Retina	Absorbs light; stores vitamin A; forms impulses which are transmitted to brain
Optic nerve	Transmits impulses to the brain
Choroid	Absorbs stray light; nourishes retina

The accessory structures of the eye include the eyebrows, eyelids, eyelashes, conjunctiva, and lachrymal apparatus. These structures have several functions, including protecting the anterior portion of the eye, preventing the entry of foreign particles, and keeping the eyeball moist.

Do the **eyes grow** like other organs?

Unlike most other organs, the eyes do not grow very much from infancy to adulthood. The average diameter of the eyeball is about 0.68 inches (17 millimeters) at birth and 0.84 inches (21 millimeters) in adulthood. However, since new lens fibers are produced throughout life, the thickness of the lens varies with age. At birth, the thickness measures from 0.14 inches (3.5 millimeters) to 0.16 inches (4 millimeters), and at age 95 it may be 0.19 inches (4.75 millimeters) to 0.20 inches (5 millimeters) thick.

What determines **eye color**?

Variations in eye color range from light blue to dark brown and are inherited. Eye color is chiefly determined by the amount and distribution of melanin within the irises. If melanin is present only in the epithelial cells that cover the posterior surface of the iris, the iris appears blue. When this condition exists together with denser-than-usual tissue within the body of the iris, the eye color looks gray. When melanin is present within the body of the iris, as well as the epithelial covering, the iris appears brown.

What causes an individual to have **eyes** with **different colors**?

Heterochromia is a condition in which one iris is a different color from the other iris. This condition, relatively rare in humans, is usually inherited though it may be caused by disease or injury. In some cases, one part of one iris is a different color from the rest of the iris, a condition known as partial or sectoral heterochromia.

What **part of the eye** is known as the **"white of the eye"**?

The sclera, or tough outer coat, is the "white of the eye."

What are the **floaters** that move around on the eye?

Floaters are semi-transparent specks perceived to be floating in the field of vision. Some originate with red blood cells that have leaked out of the retina. The blood cells swell into spheres, some forming strings, and float around the areas of the retina. Others are shadows caused by the microscopic structures in the vitreous humor, a jellylike substructure located behind the retina. A sudden appearance of a cloud of dark floaters, if accompanied by bright light flashes, could indicate retinal detachment.

How **frequently** are the **eye muscles used**?

Eye muscles may move up to 100,000 times in a 24-hour period. To give the legs the equivalent exercise would require 50 miles of walking. The eye muscles contain a special form of very rapidly contracting myosin so that they can move rapidly but not fatigue.

What are the **two layers** of the **retina**?

The two layers that comprise the retina are an outer pigmented layer called the pigment epithelium, which adheres to the choroid, and an inner layer of nerve tissue

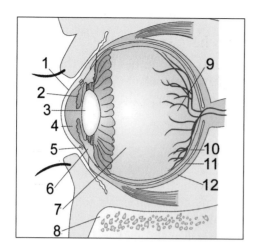

The structures of the human eye. 1 = Cornea; 2 = Iris; 3 = Lens; 4 = Anterior chamber; 5 = Canal of Schlemm; 6 = Ciliary body; 7 = Posterior chamber; 8 = Bony orbit; 9 = Macula lutea; 10 = Retina; 11 = Uvea; and 12 = Sclera. (From Pilliterri, Adele. *Maternal and Child Nursing*. 4th ed. Philadelphia: Lippincott, Williams & Wilkins, 2003.)

called the sensory (or neural) retina. The inner layer of nerve tissue consists of three separate layers of neurons. The first and closest to the choroid is a layer of sensory receptors, the photoreceptors cells called rods and cones, and various other neurons. Next is a layer of bipolar neurons, the nerve cells that receive impulses generated by the rod and cone cells. The third or inner layer consists of ganglionic neurons attached directly to the optic nerve.

Why does **diabetes** cause **blindness**?

Diabetic retinopathy is the major cause of blindness in the United States among adults ages 20 to 65. The high blood sugar level in diabetes weakens blood vessel walls in the retina and choroids, which increases susceptibility to hemorrhaging, scarring, and retinal detachment.

What is the difference in the functions of the **rods** and **cones** found in the eyes?

Rods and cones are photoreceptor cells that convert light first into chemical energy and then into electrical energy for transmission to the vision centers of the brain via the optic nerve. Rods are specialized for vision in dim light; they cannot detect color, but they are the first receptors to detect movement and register shapes. There are about 125 million rods in a human eye. They contain a pigment called rhodopsin. Cones provide acute vision, functioning best in bright daylight. They allow us to see colors and fine detail. Cones are divided into three different types that contain cyanolabe, chlorolabe, or erythrolabe. These photopigments absorb wavelengths in the short (blue), middle (green), and long (red) ranges, respectively. There are about seven million cones in each eye.

How **long** does it take for a person to **adapt** to **dim light**?

Rods give us vision in dim light but not in color and not with sharp detail. They are hundreds of times more sensitive to light than cones, letting us detect shape and movement in dim light. This type of photoreceptor takes about 15 minutes to fully adapt to very dim light.

Is **night blindness** dangerous or serious?

Night blindness is a condition in which the rods in the retina are seriously damaged due to vitamin A deficiency. This results in the inability to drive safely at night. Vit-

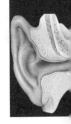

amin A supplements can reverse this condition if administered before degeneration of the rods.

What are the **different types** of **cone cells** in humans?

Color perception depends on cones. Humans have three types of cones: blue, green, and red. Each contains a slightly different photopigment. Although the retinal portion of the pigment molecule is the same as in rhodopsin, the opsin protein differs slightly in each type of photoreceptor. Each type of cone responds to light within a range of wavelengths but is named for the ability of its pigment to absorb a wavelength more strongly than the other cones. Red light, for example, can be absorbed by all three types of cones, but those cones most sensitive to red act as red receptors. By comparing the relative responses of the three types of cones, the brain can detect colors of intermediate wavelengths. There will always be a dominant cone sending the electrical color-coded impulse to the brain, but the other two color cones will also be stimulated to some degree, even if it is a faint spark. These various and unlimited combinations are what make possible the million of shades of color we see.

What are the **three different cone pigments**?

The three different cone pigments are erythrolabe, chlorolabe, and cyanolabe. Erythrolabe is most sensitive to red light waves; chlorolabe is most sensitive to green light waves; and cyanolabe is most sensitive to blue light waves.

Are the three types of **color cones present** in equal **quantities**?

In an individual with normal vision, the cone population consists of 16 percent blue cones, 10 percent green cones, and 74 percent red cones. Although their sensitivities overlap, each type is most sensitive to a specific portion of the visual spectrum.

What causes **color blindness**?

Color blindness is the inability to perceive one or more colors; this may involve complete or partial loss of color perception. Most forms of color blindness occur more frequently in males as an X-linked genetic disorder. Color blindness is actually a collection of several abnormalities of color vision. The most common form is a red-green blindness, which affects about eight percent of the male population in the United States. Red blindness is the inability to see red as a distinct color, while green blindness is the inability to see green. A rare form of color blindness is the inability to see the color blue.

How fast do photoreceptors respond?

It takes 0.002 seconds for the brain to recognize an object after its light first enters the eyes.

What are phosphenes?

If the eyes are shut tightly, the lights seen are phosphenes. Technically, the luminous impressions are due to the excitation of the retina caused by pressure on the eyeball.

What is the blind spot of the eye?

The area where the optic nerve passes out of the eyeball, the optic disc, is known as the "blind spot" because it lacks rods and cones. Images falling on the blind spot cannot be perceived.

What structure of the eye produces tears?

The lachrymal gland, which is part of the larger lachrymal apparatus, produces tears that flow over the anterior surface of the eye. Most of this fluid evaporates, but excess amounts are collected in small ducts in the corner of the eye. Tears lubricate and cleanse the eye. In addition, tears contain lysozyme, an enzyme that is capable of destroying certain kinds of bacteria and helps fight eye infections.

How often does a person blink?

The rate of blinking varies, but on average the eye blinks once every five seconds (12 blinks per minute). Assuming the average person sleeps eight hours a day and is awake for 16 hours, he or she would blink 11,520 times a day, or 4,204,800 times a year. The average blink of the human eye lasts about 0.05 seconds.

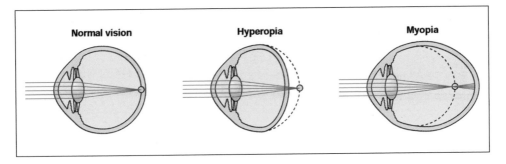

| Normal vision | Hyperopia | Myopia |

The shape of the eye and the lens determines how light focuses on the retina. (From Willis, M.C. *Medical Terminology: A Programmed Learning Approach to the Language of Health Care*. Baltimore: Lippincott, Williams & Wilkins, 2002.)

What is **nearsightedness**?

Nearsightedness, or myopia, is the ability to see close objects but not distant ones. It is a defect of the eye in which the focal point is too near the lens and the image is focused in front of the retina when looking at distant objects. This condition is corrected by concave lenses (eyeglasses or contact lenses) that diffuse the light rays coming to the eyes so that when the light is focused by the eyes it reaches the proper spot on the retinas.

What is **farsightedness**?

Farsightedness, or hyperopia, is the ability to see distant objects but not close ones. It is a disorder in which the focal point is too far from the lens, and the image is focused "behind" the retina when looking at a close object. In this condition, the lens must thicken to bring somewhat distant objects into focus. Farsightedness is corrected by a convex lens that causes light rays to converge as they approach the eye to focus on the retina.

What is **astigmatism**?

Astigmatism is an irregularity in the curvature of the cornea or lens that causes light traveling in different planes to be focused differently. The normal cornea or lens has a spherical curvature like the inside of a ball. In astigmatism, the cornea or lens has an elliptical curvature, like the inside of a spoon. As a result, some portions of an image are in focus on the retina, while other portions are blurred and vision is distorted.

What causes **double vision**?

Double vision (diplopia) is seeing two images of one object. In order for the brain to receive a clear image, both eyes must move in unison. This condition may result from weakness in one or more of the muscles that control eye movements. When this happens, the good eye focuses on a seen object, but the affected eye is focused elsewhere. Other causes include fatigue, alcohol intoxication, multiple sclerosis, or trauma. The sudden and ongoing appearance of double vision may indicate a serious disorder of the brain or nervous system.

What is **strabismus**?

Strabismus is a lack of coordination between the eyes. As a result, the eyes look in different directions and do not focus simultaneously on a single point. When the two eyes fail to focus on the same image, the brain may learn to ignore the input from one eye. If this continues, the eye that the brain ignores will never see well. This loss of vision is called amblyopia. Treatment for strabismus includes exercises and other strategies to strengthen the weakened eye muscles and realign the eyes. Glasses may also be prescribed. Surgery may be required to realign the eye muscles if strengthening techniques are unsuccessful.

What is the **most common cause of blindness** in the United States?

A person is considered legally blind in the United States when his or her best corrected vision is 20/200 or worse. Cataracts are the most common cause of blindness in the United States. This is a condition in which clouding of the lens occurs as a result of advancing age, infection, trauma, or excessive exposure to ultraviolet (UV) radiation.

How wide an angle is covered by the average person's **field of vision**?

The average field of vision is about 200 degrees.

Who invented **bifocal lenses**?

The original bifocal lens was invented in 1784 by Benjamin Franklin (1706–1790). At that time, the two lenses were joined in a metallic frame. In 1909 J. L. Borsch welded the two lenses together. One-part bifocal lenses were developed in 1910 by researches at the Carl Zeiss Company.

What **part of the eye** was the first to be successfully **transplanted**?

The cornea was the first tissue from the eye that was successfully transplanted. On December 7, 1905, Dr. Eduard Konrad Zirm (1863–1944), who was head of medicine at Olomovic Hospital in Moravia (now part of the Czech Republic), performed the first corneal transplant. It was the first successful human-to-human transplant of any organ. Interestingly, the cornea is the only tissue in the body that can be transplanted from one person to another with little or no possibility of rejection.

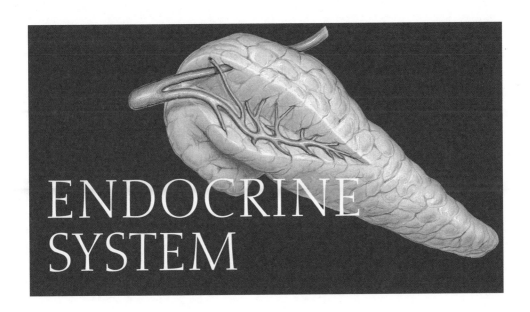

ENDOCRINE SYSTEM

INTRODUCTION

What are the **functions** of the **endocrine system**?

The endocrine system, together with the nervous system, controls and coordinates the functions of all of the human body systems. The endocrine system helps to maintain homeostasis and metabolic functions, allows the body to react to stress, and regulates growth and development, including sexual development.

What are the **similarities** between the **nervous system** and the **endocrine system**?

Both the nervous system and endocrine system are devoted to maintaining homeostasis by coordinating and regulating the activities of other cells, tissues, organs, and systems. Both systems are regulated by negative feedback mechanisms. Chemical messengers are important in both systems, although their method of transmission and release differs in the two systems.

How does the **endocrine system differ** from the **nervous system**?

Both the endocrine and nervous are regulatory systems that permit communication between cells, tissues, and organs. A major difference between the endocrine system and nervous system is the rate of response to a stimulus. In general, the nervous system responds to a stimulus very rapidly, often within a few milliseconds, while it may take the endocrine system seconds and sometimes hours or even days to offer a response. Furthermore, the chemical signals released by the nervous system typically act over very short distances (a synapse), while hormones in the endocrine system are generally carried by the blood to target organs. Finally, the effects of the

161

nervous system generally last only a brief amount of time, while those of the endocrine system are longer lasting. Examples of endocrine control are growth and reproductive ability.

What are the **organs** of the **endocrine system**?

The endocrine system consists of glands and other hormone-producing tissues. Glands are specialized cells that secrete hormones into the interstitial fluid. Hormones are then transported to the capillaries and circulated via the blood. The major endocrine glands are the pituitary, thyroid, parathyroid, pineal, and adrenal glands. Other hormone-secreting organs are the central nervous system (hypothalamus), kidneys, heart, pancreas, thymus, ovaries, and testes. Some organs, such as the pancreas, secrete hormones as an endocrine function but have other functions also.

HORMONES

What are **hormones**?

Hormones are chemical messengers that are secreted by the endocrine glands into the blood. Hormones are transported via the bloodstream to reach specific cells, called target cells, in other tissues. They produce a specific effect on the activity of cells that are remotely located from their point of origin.

What are **target cells**?

Target cells are specific cells that respond to a specific hormone. Target cells have special receptors on their outer membranes that allow the individual hormones to bind to the cell. The hormones and receptors fit together much like a lock and key.

How are **hormones classified**?

Scientists classify hormones broadly into two classes: those that are soluble in water (hydrophilic) and those that are not soluble in water (hydrophobic) but are soluble in lipids. The chemical structure of hormones determines whether they are water-soluble or lipid-soluble. Water-soluble hormones include amine, peptide, and protein hormones. The steroid hormones are lipid-soluble.

What are the major **groups** of **hormones**?

The major groups of hormones are amine hormones, peptide and protein hormones, and steroid hormones. Amine hormones are relatively small molecules that are structurally similar to amino acids. Epinephrine and norepinephrine, serotonin, dopamine, the thyroid hormones, and melatonin are examples of amine hormones.

Peptide hormones and protein hormones are chains of amino acids. The peptide hormones have 3 to 49 amino acids, while the protein hormones are larger with chains of 50 to 200 or more amino acids. Examples of peptide hormones are antidiuretic hormone and oxytocin. The larger thyroid-stimulating hormone and follicle-stimulating hormone are examples of protein hormones.

Steroid hormones are derived from cholesterol. Cortisol and the reproductive hormones (androgens in males and estrogens in females) are examples of steroid hormones.

Types of Hormones Based on Chemical Structure

Chemical Class	Water- or Lipid-soluble	Examples
Eicosanoids	Lipid soluble	Prostaglandins, leukotrienes
Steroid hormones	Lipid soluble	Aldosterone, cortisol, androgens, calcitrol, testosterone, estrogens, progesterone
Thyroid hormones	Lipid soluble	T_3 (triiodothyronine) and T_4 (thyroxine)
Amines	Water soluble	Epinephrine and norepinephrine, melatonin, histamine
Peptides and proteins	Water soluble	All hypothalamic releasing and inhibiting hormones, oxytocin, antidiuretic hormone; human growth hormone, thyroid-stimulating hormone (TSH), follicle-stimulating hormone (FSH), luteinizing hormone, prolactin, melanocyte-stimulating hormone; insulin, glucagons, somatostatin, pancreatic polypeptide; parathyroid hormone, calcitonin, gastrin, secretin, cholecystokinin, glucose-dependent insulinotropic peptide, leptin

163

Which **endocrine glands** produce which **hormones**?

Each endocrine gland produces specific hormones, as explained in the below table.

Endocrine Glands and Their Hormones

Gland	Hormone(s) Produced
Anterior pituitary	Thyroid-stimulating hormone (TSH); adrenocorticotropic hormone (ACTH); follicle-stimulating hormone (FSH); luteinizing hormone (LH); prolactin (PRL); growth hormone (GH); melanocyte-stimulating hormone (MSH)
Posterior pituitary	Antidiuretic hormone (ADH); oxytocin
Thyroid	Thyroxine (T_4); triiodothyronine (T_3); calcitonin (CT)
Parathyroid	Parathyroid hormone (PTH)
Pineal	Melatonin
Adrenal (cortex)	Mineralocorticoids, primarily aldosterone; Glucocorticoids, mainly cortisol (hydrocortisone); corticosterone; cortisone
Adrenal (medula)	Epinephrine (E); norepinephrine (NE)
Pancreas	Insulin; glucagon
Thymus	Thymosins
Ovaries	Estrogens; progesterone
Testes	Androgens, mainly testosterone

How do **paracrine hormones** differ from **circulating hormones**?

Local hormones become active without first entering the bloodstream. They act locally on the same cell that secreted them or on neighboring cells. Local secretion of pro-inflammatory factors increases extravasation from blood vessels to produce local edema and flare responses. Circulating hormones are more prevalent than local hormones. Once secreted, they enter the bloodstream to be transported to their target cells.

How long does a **hormone** remain **active** once it is released into the circulatory system?

Hormones that circulate freely in the blood remain functional for less than one hour. Some hormones are functional for as little as two minutes. A hormone becomes inactivated when it diffuses out of the bloodstream and binds to receptors in target tissues or is absorbed and broken down by cells of the liver or kidneys. Enzymes in the plasma or interstitial fluids that break down hormones also cause them to become inactivated. Other hormones (e.g., renin) are activated by enzymes that cleave the active portion from a larger circulating precursor molecule.

Where are **hormone receptors located** within a cell?

Hormone receptors are located either on the surface of the cell membrane or inside the cell. Water-soluble hormones are not able to diffuse through the plasma mem-

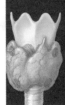

Do hormones affect behavior?

Endocrine functions and hormones interact with every other organ system in the human body. Individuals whose hormone levels are abnormal, either due to oversecretion or undersecretion of a particular hormone, will show signs of abnormal behavior and illness. Children whose sex hormones are produced at an early age, for example, may demonstrate aggressive and assertive behavior in addition to the physical characteristics of maturation. In adults, changes in hormonal levels may have significant effects on intellectual capabilities, memory, learning, and emotional states.

brane easily. Therefore, the receptors for these hormones are located on the surface of the cell. In contrast, lipid-soluble hormones are able to easily penetrate the cell membrane. The receptors for the lipid-soluble hormones are often located inside the cell.

What is the **hormonal response** to **stress**?

The stress response, also known as the general adaptation syndrome (GAS), has three basic phases: 1) the alarm phase, 2) the resistance phase, and 3) the exhaustion phase. The alarm phase is an immediate reaction to stress. Epinephrine is the dominant hormone of the alarm phase. It is released in conjunction with the sympathetic nervous system and produces the "fight or flight" response. Nonessential body functions such as digestive, urinary, and reproductive activities are inhibited.

The resistance phase follows the alarm phase if the stress lasts more than several hours. Glucocorticoids are the dominant hormones of the resistance phase. Endocrine secretions coordinate three integrated actions to maintain adequate levels of glucose in the blood. They are: 1) the mobilization of lipid and protein reserves, 2) the conservation of glucose for neural tissues, and 3) the synthesis and release of glucose by the liver.

If the body does not overcome the stress during the resistance phase, the exhaustion phase begins. Prolonged exposure to high levels of hormones involved in the resistance phase leads to the collapse of vital organ systems. Unless there is successful intervention and it can be reversed, the failure of an organ system will be fatal.

How does **aging** affect the **endocrine system**?

Most endocrine glands continue to function and secrete hormones throughout an individual's lifetime. The most noticeable change in hormonal output is in the reproductive hormones. The ovaries decrease in size and no longer respond to FSH and LH, resulting in a decrease in the output of estrogens. Although the hormonal levels of other hormones may not change with aging and remain within normal

165

limits, some endocrine tissues become less sensitive to stimulation. For example, elderly people may not produce as much insulin after a carbohydrate-rich meal is eaten. It has been suggested that the decrease in function of the immune system is a result of the reduced size of the thymus gland.

PITUITARY GLAND

Where is the pituitary gland located?

The pituitary gland is located directly below the hypothalamus in the sella turcica ("Turkish saddle"), a depression in the sphenoid bone. It is protected on three sides by the bones of the skull and on the top by a tough membrane called the diaphragma sellae.

How large is the pituitary gland?

The pituitary gland is about the size of a plump lima bean. It measures 0.39 inches (1 centimeter) long, 0.39 to 0.59 inches (1 to 1.5 centimeters) wide, and 0.12 inches (0.5 centimeters) thick.

What are the differences between the two regions of the pituitary gland?

The pituitary gland is divided into an anterior lobe (or adenohypophysis) and a posterior lobe. The anterior lobe is the larger section of the pituitary, accounting for 75 percent of the total weight of the gland. The anterior lobe contains endocrine secretory cells, which produce and secrete hormones directly into the circulatory system via an extensive capillary network that surrounds the region. The posterior lobe (or neurohypophysis) does not manufacture any hormones. It contains the axons from two different groups of hypothalamic neurons. Hormones produced in the hypothalamus are transported from the hypothalamus to the posterior pituitary within the axons.

How many different hormones are secreted by the pituitary gland?

Nine different peptide hormones are released by the pituitary gland. Seven are produced by the anterior pituitary gland and two are secreted by the posterior pituitary gland. The hormones of the anterior pituitary gland are thyroid-stimulating hormone (TSH), adrenocorticotropic hormone (ACTH), follicle-stimulating hormone

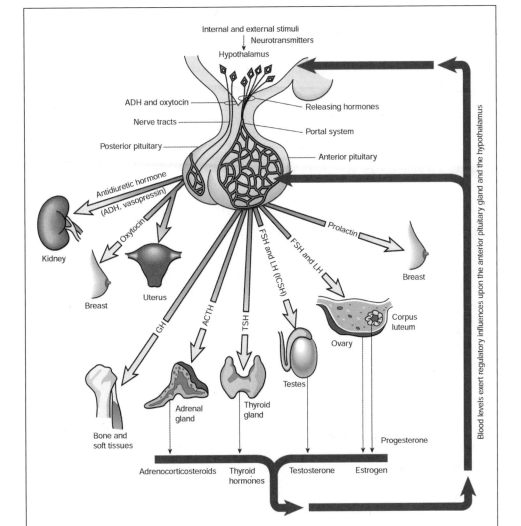

The pituitary gland secretes hormones that affect organs throughout the body. (From Smeltzer, S.C., and Bare, B.G. *Textbook of Medical-Surgical Nursing.* 9th Ed. Philadelphia: Lippincott, Williams & Wilkins, 2000.)

(FSH), luteinizing hormone (LH), prolactin (PRL), growth hormone (GH), and melanocyte-stimulating hormone (MSH). The hormones of the posterior pituitary gland are antidiuretic hormone (ADH) and oxytocin.

Which pituitary gland hormones are **trophic hormones**?

Trophic (from the Greek *trophikos,* meaning "turning toward" or "to change") hormones are hormones that regulate the production of other hormones by different endocrine glands. These hormones "turn on," or activate, the target endocrine glands. The trophic hormones are thyroid-stimulating hormone (TSH), adrenocorticotropic hormone (ACTH), follicle-stimulating hormone (FSH), and luteinizing hormone (LH).

What are the **targets** of the **trophic hormones**?

The target of each of the trophic hormones is another endocrine gland.

Pituitary Tropic Hormones and Their Targets

Hormone	Target
Thyroid-stimulating hormone (TSH)	Thyroid gland
Adrenocorticotropic hormone (ACTH)	Adrenal glands
Follicle-stimulating hormone (FSH)	Gonads
Luteinizing hormone (LH)	Gonads

How do **follicle-stimulating hormone (FSH)** and **luteinizing hormone (LH)** have different actions in **males** and **females**?

Both follicle-stimulating hormone and luteinizing hormone are gonadotropic hormones. In females, FSH promotes the growth and development of follicle cells in ovaries. Follicle cells surround a developing oocyte. In response to FSH they grow and develop to the point that one ruptures and expels an ovum to be fertilized. In males, FSH stimulates the production of sperm in the testes.

Luteinizing hormone (LH) induces ovulation, the release of an egg by the ovary, in females. It also stimulates the secretion of estrogen and the progestins, such as progesterone. In males, LH stimulates the production and secretion of androgens, the male sex hormones, including testosterone.

Which medical condition is caused by **low production** of **gonadotropins**?

Hypogonadism is caused by abnormally low production of gonadotropins. Children with this condition will not undergo sexual maturation. Adults with hypogonadism cannot produce functional sperm or oocytes (eggs).

What are the functions of **prolactin**?

Prolactin has two major functions in females. First, it works together with other hormones to stimulate the development of the ducts in the mammary glands. Secondly, it stimulates the production of milk after childbirth. Most researchers believe that prolactin has no effect in males, while some believe it may help regulate androgen production.

Which **cells and tissues** are most affected by **human growth hormone**?

Human growth hormone, sometimes called just growth hormone (GH), affects all parts of the body associated with growth. The skeletal muscles and cartilage cells are especially sensitive to the levels of growth hormone. One of the most direct effects of growth hormone is to maintain the epiphyseal plates of the long bones, where growth takes place.

Which conditions result from **disorders** of the **human growth hormone**?

A deficiency of human growth hormone in children causes growth at a slower than normal rate during puberty. Slow epiphyseal growth results in short stature and larger than normal adipose tissue reserves. In contrast, if there is no decrease in the secretion of GH towards the end of adolescence, the individual will continue to grow to seven or even eight feet tall, resulting in gigantism. When GH is overproduced after normal growth has ceased, a condition called acromegaly (from the Greek *akros,* meaning "extremity," and *megas,* meaning "great" or "big") occurs. Although the epiphyseal discs cartilages have closed, the small bones in the head, hands, and feet continue to grow, thickening rather than lengthening.

When is **growth hormone therapy** recommended?

Growth hormone therapy involves the injection of synthetic growth hormone several times a week for a period of several years to stimulate growth. It is most effective when started before the bones have fused so that there is still the potential for growth. Not recommended for every short child, it is beneficial for children who have trouble functioning (e.g., climbing the steps of the school bus).

What is the function of **antidiuretic hormone**?

The primary function of antidiuretic hormone (ADH), or vasopressin, is to decrease the amount of urine excreted and increase the amount water absorbed by the kidneys. It plays a critical role in regulating the balance of fluids in the body.

What conditions **increase** the secretion of **ADH**?

Secretion of ADH increases in response to fluid loss, such as dehydration. Hemorrhaging causes an increase in ADH secretion in order to maintain the body's fluid balance. Very strenuous exercise, emotional or physical stress, and drugs such as nicotine or barbiturates all increase the secretion of ADH in order to decrease the amount of urine excreted.

What are the functions of **oxytocin**?

Oxytocin (from the Greek *oxy,* meaning "quick," and *tokos,* meaning "childbirth") stimulates contractions of the smooth muscle tissue in the wall of the uterus during childbirth. Prior to the late stages of pregnancy, the uterus is relatively insensitive to oxytocin. As the time of delivery approaches, the muscles become sensitive to increased secretion of oxytocin.

After delivery, oxytocin stimulates the ejection of milk from the mammary glands. The suckling of an infant stimulates the nerve cells in the brain (the hypothalamus) to release oxytocin. Once oxytocin is secreted into the circulatory system, special cells contract and release milk into collecting chambers from which the milk is released. This reflex is known as the milk let-down reflex.

What external factors can **influence** the **milk let-down reflex**?

The milk let-down reflex may be controlled by a factor that affects the hypothalamus. Anxiety and stress can prevent the flow of milk. Some mothers learn to associate a baby's crying with suckling. These women may experience the milk let-down reflex as soon as they hear their baby crying.

THYROID AND PARATHYROID GLANDS

What are the physical **characteristics** of the **thyroid gland**?

The thyroid gland is located in the neck, anterior to the trachea, just below the larynx (the voice box). It has two lobes connected by a slender bridge of tissue called the isthmus. The average weight of the thyroid gland is 1.2 ounces (34 grams). An extensive, complex blood supply gives the thyroid a deep red color.

What are the **two types of cells** in the **thyroid gland**?

The main type of cells in the thyroid are follicular cells. These cells produce T_4 (thyroxine) and T_3 (triiodothyronine). Parafollicular cells are also found between the follicles in the thyroid gland. These cells are not as numerous as follicular cells. They produce the hormone calcitonin.

Where are the **thyroid hormones stored** in the thyroid gland?

Thyroid hormones are stored in the spherical sacs called follicles. The thyroid follicles, microscopic spherical sacs, are composed of a single layer of cuboidal epithelium tissue. The thyroid hormones are stored in a gelatinous colloid.

What is a **unique characteristic** of the **thyroid gland**?

The thyroid gland is the only gland that is able to store its secretions outside its principal cells. In addition, the stored form of the hormones is different from the actual hormone that is secreted into the blood system. Enzymes break down the stored chemical prior to its release into the blood.

How does **triiodothyronine (T₃),** differ from **thyroxine (T₄)?**

Thyroxine, or T₄, also called tetraiodothyronine, contains four atoms of iodine. Triiodothyronine, or T₃, contains only three atoms of iodine. The more common hormone is T₄, which accounts for nearly 90 percent of the secretions from the thyroid. The amount of T₃ in the body is concentrated and very effective. Both hormones have similar functions. Enzymes in the liver can convert T₄ to T₃.

What are the **functions** of the **thyroid hormones?**

Thyroid hormones affect almost every cell in the body. Some important effects of thyroid hormones on various cells and organ systems are:

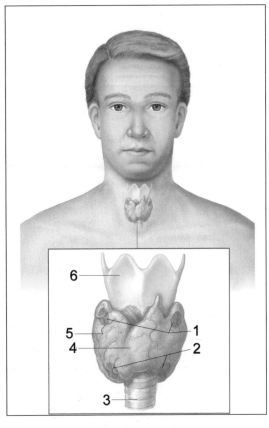

The thyroid and parathyroid glands are located in the neck: 1 = Superior parathyroids; 2 = Inferior parathyroids; 3 = Trachea; 4 = Isthmus; 5 = Thyroid gland; and 6 = Thyroid cartilage. *Anatomical Chart Co.*

- Increases body metabolism by increasing the rate at which cells use oxygen and food to produce energy
- Causes the cardiovascular system to be more sensitive to sympathetic nervous activity
- Increases heart rate and force of contraction of heart muscle
- Maintains normal sensitivity of respiratory centers to changes in oxygen and carbon dioxide concentrations
- Stimulates the formation of red blood cells to enhance oxygen delivery
- Stimulates the activity of other endocrine tissues
- Ensures proper skeletal development in children

What is the **calorigenic effect?**

The calorigenic effect is a result of the increased metabolic rate and increased oxygen consumption by cells. When the metabolic rate increases, more heat is generated and body temperature rises. Since energy use is measured in calories, this is known as the calorigenic effect.

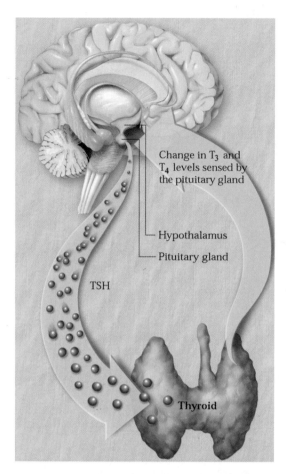

Change in T$_3$ and T$_4$ levels sensed by the pituitary gland

Hypothalamus
Pituitary gland

TSH

Thyroid

The thyroid produces the hormones T3 and T4. When these levels are too low, the pituitary gland creates more of the hormone TSH, and when T3 and T4 are too high, the pituitary makes less TSH. Doctors can measure TSH to find out the health of the thyroid. *Anatomical Chart Co.*

Why is it **important** to include **iodized salt** in one's **diet**?

Iodized salt provides an adequate supply of iodine in the diet, a necessary element for the production of thyroid hormones. A quarter teaspoon of salt (1.5 grams) provides 67 micrograms of iodine, which is about half of the U.S. Recommended Daily Allowance of 150 micrograms per day for adults.

Why is **calcitonin** important?

Calcitonin helps regulate the calcium in the blood. When the concentration of calcium ions rises in the blood, there is an increase in the secretion of calcitonin to reduce the level of calcium in the blood. When the calcium concentration falls and returns to normal, the secretion of calcitonin ceases. Calcitonin is important because it stimulates bone growth, especially in children.

What medical condition is caused by an **overactive thyroid gland**?

Hyperthyroidism is the clinical term for an overactive thyroid gland. Patients with hyperthyroidism produce excessive quantities of the thyroid hormones thyroxine and triiodothyronine, which speed up the metabolic processes and functions of the body. Symptoms of hyperthyroidism include sudden weight loss without dieting, rapid and racing heartbeat, nervousness, irritability, tremors, increased perspiration, more frequent bowel movements, and changes in menstrual patterns.

What is the **most common cause** of **hyperthyroidism**?

The most common cause of hyperthyroidism is Graves' disease. Graves' disease is an autoimmune disorder in which antibodies produced by the immune system stimulate the thyroid to produce too much thyroxine. In Graves' disease, antibodies mistakenly attack the thyroid gland and occasionally the tissue behind the eyes and the

skin of the lower legs. Some people with Graves' disease have exophthalmos, a bulging of the eyes associated with Graves' disease. Treatment options include antithyroid drugs, radioactive iodine, or surgery. Radioactive iodine is the most common treatment for hyperthyroidism. Some thyroid cells will absorb the radioactive iodine. After a period of several weeks, the cells that took up the radioactive iodine will shrink and thyroid hormone levels will return to normal.

What is a **goiter**?

A goiter is an enlargement of the thyroid gland. Goiters are often associated with an underactive thyroid (hypothyroidism), although other conditions may also produce a goiter.

What are some **common symptoms** of **hypothyroidism**?

Hypothyroidism is caused by a lack of thyroid hormone. In general, the metabolism is slower and fatigue is common. Other symptoms include feeling cold, having dry skin and hair, constipation, weight gain, muscle cramps, and

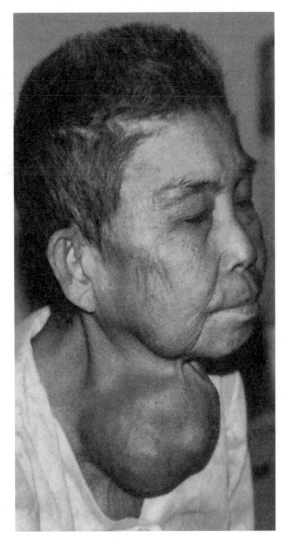

A man suffering from a goiter, or enlarged thyroid. (From Rubin, E., *Essential Pathology*. 3rd Ed. Philadelphia: Lippincott, Williams & Wilkins, 2000.)

increased menstrual flow. These symptoms may be cured by administering a synthetic form of the thyroid hormone thyroxine.

Where are the **parathyroid glands located**?

The parathyroid glands are embedded in the posterior surface of the thyroid gland. There are usually four parathyroid glands—two pairs—on each side of the thyroid. The parathyroid glands are tiny, pea-shaped glands weighing only a total of 0.06 ounces (1.6 grams) together. Each measures approximately 0.1 to 0.3 inches (3 to 8 millimeters) in length, 0.07 to 0.2 inches (2 to 5 millimeters) in width, and 0.05 inches (1.5 millimeters) in depth.

What is the **function** of the **parathyroid glands**?

The parathyroid glands secrete parathyroid hormone (PTH). The main function of PTH is to regulate the levels of calcium and phosphate in the blood.

How does **parathyroid hormone** increase the level of **calcium** in the blood?

There are four ways that parathyroid hormone (PTH) increases the level of calcium in the blood.

1. PTH stimulates osteoclasts to break down bone tissue and release calcium ions from the bone.
2. PTH inhibits osteoblasts to reduce the rate of calcium deposition in bone.
3. PTH enhances the absorption of calcium and phosphate from the small intestine in conjunction with the secretion of calcitriol by the kidneys.
4. PTH promotes the reabsorption of calcium at the kidneys, reducing the amount of calcium excreted in urine.

What is the **relationship** between **parathyroid hormone (PTH)** and **calcitonin**?

The thyroid gland secretes calcitonin when the calcium level in the blood is elevated. When the blood calcium level drops, the parathyroid glands increase the secretion of parathyroid hormone until the blood calcium level increases to the normal value. Homeostasis of blood calcium levels is maintained through the interaction of calcitonin and PTH.

Are all four **parathyroid glands** necessary to maintain **homeostasis**?

No, not all of the parathyroid glands are needed to maintain homeostasis. The secretion of even a portion of one gland can maintain normal calcium concentrations. When calcium concentration levels are abnormally high, though, a condition called hyperparathyroidism results. This is typically caused by a tumor, and surgical removal of the overactive tissue will often correct the imbalance.

What are the most common **causes** of **hypoparathyroidism**?

Hypoparathyroidism is most often the result of damage to or removal of the parathyroid glands during surgery or when a tumor is on the parathyroid gland. It is suspected that prior to the discovery of the parathyroid glands, many patients died following thyroid surgery since the parathyroid glands were accidentally removed, resulting in abnormally low levels of calcium.

ADRENAL GLANDS

What are the physical **characteristics** of the **adrenal glands**?

The adrenal (from the Latin, meaning "upon the kidneys") glands sit on the superior tip of each kidney. Each adrenal gland weighs approximately 0.19 ounces (7.5 grams). The glands are yellow in color and have a pyramid shape. Each adrenal gland has two sections that may almost be considered as separate glands. The inner portion is the adrenal medulla (from the Latin *marrow,* meaning "inside"). The outer portion, which surrounds the adrenal medulla, is the adrenal cortex (from the Latin, meaning "bark," because its appearance is similar to the outer covering of a tree). The adrenal cortex is the larger part of the adrenal glands, accounting for nearly 90 percent of the gland by weight.

How many different types of **hormones** are secreted by the **adrenal cortex**?

The adrenal cortex secretes more than two dozen different steroid hormones called the adrenocortical steroids, or simply corticosteroids. The adrenal cortex is divided into three major zones or regions each of which secretes a different type of corticosteroid. The outer region is the zona glomerulosa, which produces mineralocorticoids. The middle zone is the zona fasciculata, which accounts for the bulk of the cortical volume and produces glucocorticoids. The innermost, and smallest region, is the zona reticularis, which produces small quantities of the sex hormones.

What are the **functions** of the **corticosteroids**?

The corticosteroids are vital for life and well-being. Each of the corticosteroids serves a unique purpose.

Corticosteroids and Their Functions

Hormone	Target	Effects
Mineralocorticoids	Kidneys	Increases reabsorption of sodium ions and water from the urine; stimulates loss of potassium ions through excretion of urine
Glucocorticoids	Most cells	Releases amino acids from skeletal muscles, lipids from adipose tissues; promotes liver glycogen and glucose formation; promotes peripheral utilization of lipids; anti-inflammatory effects
Androgens		Promotes growth of pubic hair in boys and girls; in adult women, promotes muscle mass, blood cell formation, and supports the libido; in adult men, adrenal androgens are less signficant because androgens are released primarily from the gonads

175

Which is the **main mineralocorticoid** hormone?

Mineralocorticoids are responsible for regulating the concentration of minerals in the body fluids. The main mineralocorticoid is aldosterone. It increases reabsorption of sodium from the urine into the blood and it stimulates excretion of potassium into the urine. Aldosterone is released in response to a drop in the blood sodium concentration or blood pressure or to a rise in the blood potassium concentration.

What are the **effects** of **glucocorticoids** on the body?

Glucocorticoids have many varying effects on the body.

- The most critical effect of the glucocorticoids is the stimulation of glucose synthesis and glycogen formation, especially within the liver.
- They stimulate the release of fatty acids from adipose tissue, which can be used as an energy source.
- They decrease the effects of physical and emotional stress, such as fright, bleeding, and infection, since the additional supply of glucose from the liver provides tissues with a ready source of ATP.
- They suppress allergic and inflammatory reactions.
- They decrease and suppress the activities of white blood cells and other components of the immune system.

What **two disorders** are associated with **abnormal glucocorticoid production**?

Addison's disease and Cushing's syndrome are both disorders caused by abnormal glu-cocorticoid production. Addison's disease occurs when the adrenal glands do not produce enough of the hormone cortisol and, in some cases, the hormone aldosterone. Common symptoms of Addison's disease are chronic, worsening fatigue, muscle weakness, loss of appetite, and weight loss. Treatment of Addison's disease involves replacing, or substituting, the hormones that the adrenal glands are not making.

Cushing's syndrome is caused by prolonged exposure of the body's tissues to high levels of the hormone cortisol. The symptoms vary, but most people have upper body obesity, a characteristic rounded "moon" face, increased fat around the neck, and thinning arms and legs. The skin, which becomes fragile and thin, bruis-

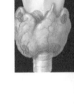

es easily and heals poorly. The symptoms of Cushing's syndrome may appear with prolonged use of prescribed glucocorticoid hormones, including prednisone.

Which are the two main **hormones** secreted by the **adrenal medulla**?

The adrenal medulla secretes epinephrine (also called adrenaline) and norepinephrine (also called noradrenaline). Epinephrine makes up 75 to 80 percent of the secretions from the adrenal medulla, with norepinephrine accounting for the remainder. These hormones are similar to those released by the sympathetic nervous system, but their effects last longer because they remain in the blood for longer periods of time.

PANCREAS

Where is the **pancreas located**?

The pancreas (from the Greek, meaning "all flesh") is located in the abdominopelvic cavity between the stomach and the small intestine. It is an elongated organ about 6 inches (12 to 15 centimeters) long.

Why is the **pancreas called** a **mixed gland**?

The pancreas is a mixed gland because it has both endocrine and exocrine functions. As an endocrine gland, it secretes hormones into the bloodstream. Only one percent of the weight of the pancreas serves as an endocrine gland. The remaining 99 percent of the gland has exocrine functions. Its functions as an exocrine gland are discussed in chapter 12 (digestion).

Which cells of the **pancreas secrete hormones**?

The pancreatic islets (islet of Langerhans) are cluster cells that secrete hormones. There are between 200,000 and 2,000,000 pancreatic islets scattered throughout the adult pancreas.

Who first described the **pancreatic islets**?

Paul Langerhans (1847–1888) was the first to provide a detailed description of microscopic pancreatic structures in the late 1860s. He noticed unique polygonal cells in the pancreas. It was not until 1893 that G.E. Laguesse discovered that the polygon-shaped cells were the endocrine cells of the pancreas that secreted insulin.

177

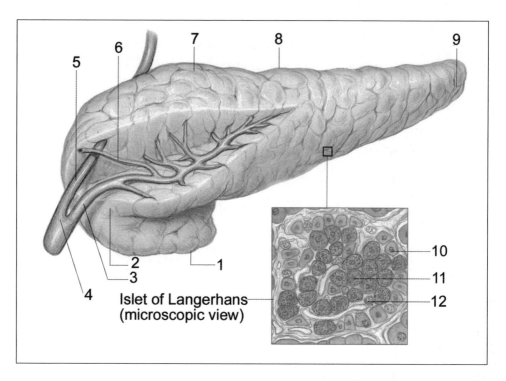

The parts of the pancreas: 1 = Uncinate process; 2 = Head; 3 = Main pancreatic duct; 4 = Hepatopancreatic ampulla; 5 = Common bile duct; 6 = Accessory pancreatic duct; 7 = Neck; 8 = Body; 9 = Tail. The Islet of Langerhans includes 10 = Alpha cell; 11 = Beta cell; 12 = Delta cell. *Anatomical Chart Co.*

How many different **types of cells** are found in the **islets of Langerhans**?

There are four different types of cells in each of the islets of Langerhans. The four groups of cells are alpha cells, beta cells, delta cells, and F cells. The two most important types of cells are alpha cells, which produce glucagon, and beta cells, which produce insulin.

What is the **function** of **glucagon**?

Glucagon is secreted when the blood glucose levels fall below normal values. Glucagon stimulates the liver to convert glycogen to glucose, which causes the blood glucose level to rise. Glucagon also stimulates the production of glucose from amino acids and lactic acid in the liver. Glucagon stimulates the release of fatty acids from adipose tissue. When blood glucose levels rise, the secretion of glucagons decreases as part of the negative feedback system.

Who **discovered insulin**?

Insulin was discovered by Frederick Banting (1891–1941), John Macleod (1876–1935), and Charles Best (1899–1978). Although earlier researchers had suspected that the pancreas secreted a substance that controlled the metabolism of sugar, it was not proved until 1922, when Banting, Macleod, and Best announced their dis-

covery. The Nobel Prize in Physiology or Medicine in 1923 was awarded to Banting and Macleod. Banting shared his half of the prize with Best, while Macleod shared his half of the prize with James Bertram Collip (1892–1965), who had also collaborated with the team.

What is the **function** of **insulin**?

Insulin is secreted when blood glucose levels rise above normal values. One of the most important effects of insulin is to facilitate the transport of glucose across plasma membranes, allowing the diffusion of glucose from blood into most body cells. It also stimulates the production of glycogen from glucose. The glucose is then stored in the liver to be released when blood glucose levels drop.

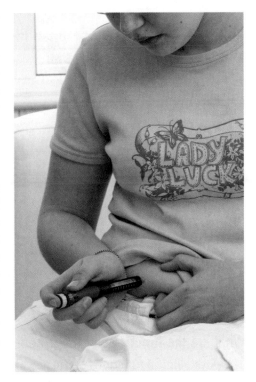

The availability of synthesized insulin has allowed a vastly improved quality of life for diabetics. © iStockphoto.com/Dave Parker.

When was the **structure** of **insulin** determined?

The full structure of insulin, a peptide hormone, was determined in 1955 by Frederick Sanger (1918–). It was the first protein to have its full structure determined. Sanger won the Nobel Prize in Chemistry in 1958 for his research.

Which medical condition is caused by the body's **inability** to produce or use **insulin**?

Diabetes mellitus (from the Greek, meaning "siphon" or "to pass through," and *meli,* meaning "honey") is a disorder of the metabolism caused when the pancreas either produces little or no insulin, or when the cells do not respond appropriately to the insulin that is produced. Glucose builds up in the blood, overflows into the urine, and passes out of the body. As a result, the body does not benefit from glucose as a source of energy.

How does **diabetes insipidus** differ from **diabetes mellitus**?

Diabetes mellitus results from an inability to produce insulin, while diabetes insipidus is the result of the pituitary not releasing sufficient quantities of antidiuretic hormone (ADH). Water conservation at the kidneys is impaired and excessive amounts of urine are excreted.

PINEAL GLAND

Where is the **pineal gland located**?

The pineal gland (from the Latin *pinea,* meaning "pinecone") is a small gland located in the midbrain at the posterior end of the third ventricle.

What are the **functions** of the **pineal gland**?

The physiological functions of the pineal are unclear. It secretes the hormone melatonin, which appears to be associated with circadian rhythms and setting the biological clock. Melatonin is mostly released at night when it is dark and its secretion diminishes during the day when it is light. The release of melatonin can be influenced by artifically mimicking day and night, such as with indoor lighting; also, its effects on sleep explain why it is used in some medications to induce sleep.

When was **melatonin** first **discovered**?

Melatonin was discovered by Aaron B. Lerner (1920–) in 1958. Richard J. Wurtman did much of the pioneering research on the benefits of melatonin.

What is **seasonal affective disorder (SAD)**?

Seasonal affective disorder is a type of depression that affects some individuals during the winter months when there is less sunlight. One hypothesis is that since there are fewer hours of daylight during the winter months, the production of melatonin is affected, resulting in physical ailments such as drowsiness and lethargy. Additional symptoms of SAD include a craving for carbohydrates, increased appetite, weight gain, and mood swings. Many researchers believe light therapy is an effective treatment for SAD. Light therapy, also called phototherapy, involves sitting near a specially designed light box that produces a strong light. Most light boxes emit a light of 2,500 to 10,000 lux, which is between the average living room lighting of 100 lux and a bright sunny day of about 100,000 lux.

What **hormone** can be used to **overcome jet lag**?

Jet lag occurs when an individual's biological clock is out of sync with local time. As a general rule it takes about a day for each hour of time zone change to recover from jet lag. Melatonin, available as a dietary supplement, is sometimes used to induce sleep when traveling. It is more useful when traveling east and may be taken before, during, or after traveling. It is best taken approximately five to seven hours before the usual bedtime in the old time zone. Travelers should consult their physicians before using melatonin. It is not recommended for pregnant or breast-feeding women and children.

REPRODUCTIVE ORGANS

Which **reproductive organs** secrete **hormones**?

The gonads (from the Greek *gonos,* meaning "offspring") release hormones in both males and females. In males, the testes secrete hormones, while in females the ovaries are responsible for secreting hormones.

What are the **hormones** of the **reproductive glands**?

The androgen hormone testosterone is the most important male hormone secreted by the testes. The testes also produce inhibin, which inhibits the secretion of FSH (follicle-stimulating hormone). The three major hormones secreted by the ovaries are estrogens, progestins, and relaxin.

What are the **functions** of the **sex hormones**?

Testosterone is stimulated by luteinizing hormone (LH) from the pituitary gland. It regulates the production of sperm, as well as the growth and maintenance of the male sex organs. Testosterone also stimulates the development of the male secondary sex characteristics, including growth of facial and pubic hair. It causes the deepening of the male voice by enlarging the larynx.

The estrogens are stimulated by follicle-stimulating hormone (FSH) in the pituitary. They help regulate the menstrual cycle and the development of the mammary glands and female secondary sex characteristics. Luteinizing hormone (LH) stimulates the secretion of progestins. Progesterone prepares the uterus for the arrival of a developing embryo in case fertilization occurs. It also accelerates the movement of an embryo to the uterus. Relaxin helps enlarge and soften the cervix and birth canal at the time of delivery. It causes the ligaments of the pubic symphysis to be more flexible at the time of delivery.

What are **anabolic steroids**?

Properly called anabolic-androgenic steroids, anabolic steroids are hormones that work to increase synthesis reactions, particularly in muscle. They are synthetic ver-

How are anabolic steroids abused?

Anabolic steroids are often abused by teenagers, adults, and athletes, both professional and amateur, for bodybuilding and to enhance athletic performance. Although they have clinical applications for certain medical conditions, the doses prescribed legally to treat these medical conditions are 10 to 100 times lower than the doses that are abused for performance enhancement.

sions of the primary male sex hormone testosterone. They promote the growth of skeletal muscle (anabolic effects) and the development of male sexual characteristics (androgenic effects).

OTHER SOURCES OF HORMONES

When was the hormone **leptin discovered**?

Leptin is a recently discovered hormone. Jeffrey Friedman (1954–) and his colleagues published a paper in December 1994 announcing the discovery of a gene in mice and humans called obese (ob) that codes for a hormone he later named leptin, after the Greek word *leptos,* meaning "thin." Leptin is a hormone made by the body's adipose tissue that regulates food intake and energy expenditure. Individuals who lack leptin eat tremendous amounts of food and become obese.

Where is the **thymus gland located**?

The thymus gland is located in the mediastinum, generally posterior to the sternum and between the lungs. It is a double-lobed lymphoid organ well supplied with blood vessels but few nerve fibers. The outer cortex of the thymus has many lymphocytes, while the inner medulla contains fewer lymphocytes.

What is the **function** of the **thymus gland**?

The thymus gland produces several hormones, called thymosins, which stimulate the production and development of T cells. T cells play an important role in immunity (see chapter 10 on the lymphatic system).

Which **other organs** perform **endocrine functions**?

Many organs perform endocrine functions in addition to their main functions. Some of these have been discussed in detail in this chapter; others are described in the chart below.

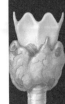

Why is the thymus called the "shrinking gland"?

The thymus is largest during infancy. It reaches its maximum effectiveness during adolescence after which time the size of the gland begins to decrease. By age 50, the thymus gland has atrophied to a fraction of its original size and is replaced by adipose tissue.

Hormones Produced by Other Organs

Organ	Hormone(s)	Effects
Intestines	Gastrin	Promotes secretion of gastric juice and increases motility in the stomach
	Secretin	Stimulates secretion of pancreatic juice and bile
	Cholecystokinin	Stimulates secretion of pancreatic juice; regulates release of bile from the gall bladder; brings about a feeling of fullness after eating
Kidneys	Erythropoietin	Stimulates red blood cell production
	Calcitrol	Stimulates calcium and phosphate absorption; stimulates calcium ion release from the bone and inhibits PTH release
Heart	Atrial natriuretic peptide (ANP)	Increases water and salt loss at kidneys; increases thirst; suppresses secretion of ADH and aldosterone
Adipose tissue	Leptin	Suppresses appetite

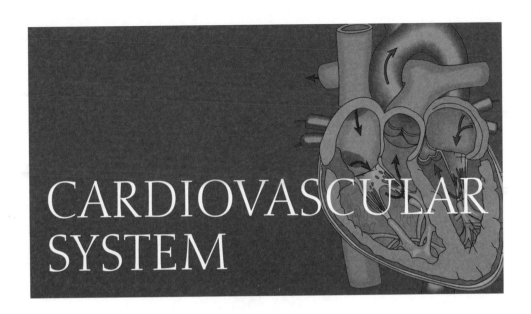

CARDIOVASCULAR SYSTEM

INTRODUCTION

What are the **functions** of the **cardiovascular system**?

The cardiovascular system provides a transport system between the heart, lungs and tissue cells. The most important function is to supply nutrients to tissues and remove waste products.

What is the difference between the **cardiovascular system** and the **circulatory system**?

The cardiovascular system refers to the heart (cardio) and blood vessels (vascular). The circulatory system is a more general term encompassing the blood, blood vessels, heart, lymph, and lymph vessels.

Which **structures** and **organs** constitute the **cardiovascular system**?

Technically speaking, the structures of the cardiovascular system are the heart and blood vessels. Blood, a connective tissue, plays a major role in the cardiovascular system and is usually discussed within the context of the cardiovascular system.

What are the some **common cardiovascular diseases**?

Cardiovascular disease is a generic term for diseases of the heart (cardio) and blood vessels (vascular). Some cardiovascular diseases are congenital (present at birth), while others are acquired later in life. Heart diseases affect the heart, arteries that supply blood to the heart muscle, or valves that ensure that blood in the heart is pumped in the correct direction. Examples of heart disease are coronary artery disease (diseases of the arteries, which supply the heart with blood), valvular heart dis-

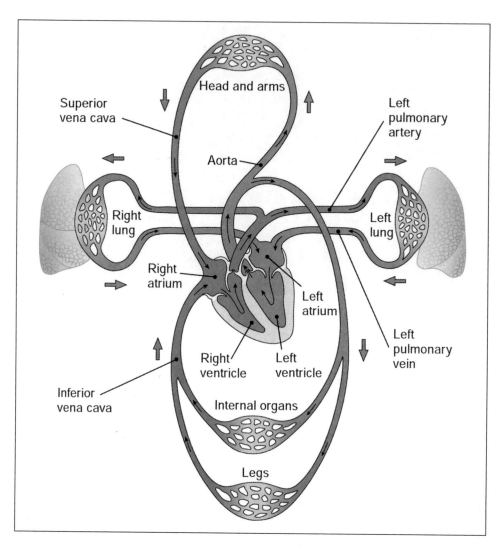

A simplified diagram of the basic cardiovascular system. (From Cohen, B. J., and Wood, D.L. *Memmler's The Human Body in Health and Disease*. 9th Ed. Philadelphia: Lippincott, Williams & Wilkins, 2000.)

ease (diseases affecting the heart valves), congenital heart disease, and heart failure. Disorders of the blood vessels include arteriosclerosis, hypertension (high blood pressure), stroke, aneurysm, venous thrombosis (formation of blood clots in a vein), and varicose veins.

How does **exercise affect** the **cardiovascular system**?

According to the American Heart Association, physical inactivity is a major risk factor for heart disease, stroke, and coronary artery disease. Regular aerobic physical activity (brisk walking, running, jogging) not only increases general fitness levels and capacity for exercise, but it also plays a role in the prevention of cardiovascular disease. Regular physical activity can also control blood lipids. Other benefits of regular aerobic physical activity include reducing high blood pressure, reducing triglyceride levels, and increasing HDL ("good") cholesterol levels. Healthy individuals should include at least 30 minutes of moderate-to-vigorous aerobic physical activity on most days of the week. The 30-minute total may be divided into 10 or 15 minute sessions, but should attain 50 to 75 percent of maximum heart rate.

BLOOD

What is the **composition** of **blood**?

Blood is classified as a connective tissue because it has both fluid and solid (cellular) components. The fluid is plasma, in which plasma proteins and cells (red blood cells, white blood cells, and platelets) are suspended in the watery base.

What are the **functions** of **blood**?

The functions of blood can be divided into three general categories: transport, regulation, and protection.

Functions of Blood

Function	Examples
Transportation	Gases (oxygen and carbon dioxide), nutrients, metabolic waste
Regulation	Body temperature, normal pH, fluid volume/pressure
Protection	Against blood loss, against infection

What is the **normal pH** of blood?

The normal pH of arterial blood is 7.4, while the pH of venous blood is about 7.35. Arterial blood has a slightly higher pH because it has less carbon dioxide.

Why is **blood sticky**?

Blood is sticky because it is denser than water and about five times more viscous than water. Blood is viscous mainly due to the red blood cells. When the number of these cells increases, the blood becomes thicker and flows slower. Conversely, if the number of red blood cells decreases, blood thins and flows faster.

Where are **red blood** cells **formed**?

Red blood cell formation, or erythropoiesis, occurs in the red bone marrow located in the vertebrae, sternum, ribs, skull, scapulae, pelvis, and proximal limb bones. Red blood cells begin as large, immature cells (proeythroblasts), and over a seven-day period they change into a much smaller, mature, red blood cell that then enters the blood stream.

187

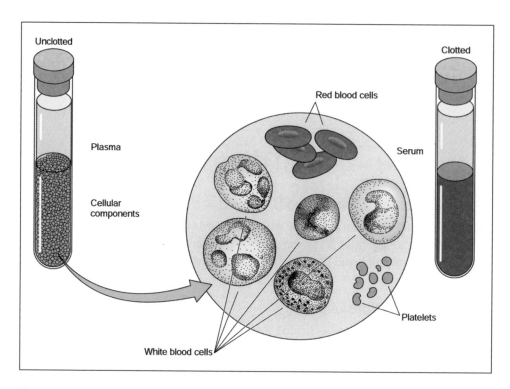

Several types of blood cells, including platelets, white and red cells, serve functions such as immunity, healing, and the transport of oxygen. (From Willis, M. C. *Medical Terminology: A Programmed Learning Approach to the Language of Health Care*. Baltimore: Lippincott, Williams & Wilkins, 2002.)

What factors can **affect** the rate of **red blood cell formation**?

The rate of red blood cell formation is stimulated by the hormone erythropoietin, which is released by the kidneys. Erythropoietin increases the rate of red blood cell division and maturation of immature red blood cells. If the blood oxygen level decreases, due to anemia, disease, or high altitude, erythropoietin is released.

How **numerous** are **red blood cells**?

Red blood cells account for one-third of all cells in the body and 99.9 percent of the cells in blood. If all the red blood cells in the human body were stacked on top of one another, it would create a tower 31,000 miles (49,890 kilometers) high.

How **thick** are **red blood cells**?

A stack of five hundred blood cells would measure only 0.04 inches (0.10 centimeters) high.

Why do **red blood cells** have such a **short life span**?

The average red blood cell lives for 120 days. Red blood cells are subject to mechanical stress as they flow through the various blood vessels in the body, creating

Why are red blood cells disc shaped?

Red blood cells are perhaps the most specialized cells in the human body. They are a biconcave (donut) shape with a thin central disc. This shape is important because the disc increases the surface-area-to-volume ratio for faster exchange of gases and it allows red blood cells to stack, one on another, as they flow through very narrow vessels. Also, since some capillaries are as narrow as 0.00015748 inches (0.004 millimeters), red blood cells can literally squeeze through narrow vessels by changing shape.

tremendous wear and tear. After about 120 days, the cell membrane ruptures and the red blood cell dies.

How **fast** are **red blood cells replaced**?

Between two million and three million red blood cells enter the blood stream each second, replacing the same number that are destroyed each second.

How does **blood transport oxygen**?

Red blood cells contain hemoglobin, which is responsible for both oxygen and carbon dioxide transport in the blood. Hemoglobin is a complex protein made of four polypeptide chains, each of which has the unique ability to bind oxygen. Amazingly enough, there are about 280 million molecules of hemoglobin in each red blood cell with the potential ability to carry a billion molecules of oxygen. Each polypeptide chain has a single molecule of heme, which has an iron ion at its center. It is the iron ion that interacts with oxygen.

What are the **major disorders** that affect **red blood cells**?

The main erythrocyte disorders are anemias in which blood has a low oxygen carrying capacity. Anemia is a symptom of an underlying disease such as lowered red blood cell count, low hemoglobin levels, or abnormal hemoglobin formation. A decrease in red blood cells may be due to hemorrhage (excessive loss of blood), iron deficiency, or reduced oxygen availability. High altitudes or some cases of pneumonia may lead to reduced oxygen availability.

What is **sickle cell anemia**?

Sickle cell anemia is an inherited disorder in which there is a mutation in one of the 287 amino acids that make up the beta chain of hemoglobin. This slight change causes abnormal folding of the hemoglobin chains so that the hemoglobin forms stiff rods. The resultant red blood cells are crescent shaped (hence the term "sickle

189

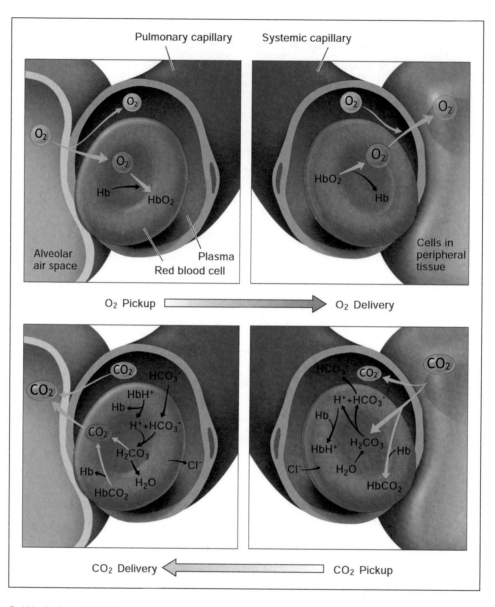

Red blood cells are designed to transport oxygen (O_2) and carbon dioxide (CO_2) through the circulatory system. (From Premkumar K. *The Massage Connection Anatomy and Physiology.* Baltimore: Lippincott, Williams & Wilkins, 2004.)

cells") when they unload oxygen or when the oxygen content of blood is reduced. These crescent shaped red blood cells are fragile and can rupture easily and form clots in small vessels.

What is **thalassemia**?

Thalassemia is an inherited genetic blood disorder that causes mild to severe anemia. The anemia in thalassemia occurs not because of a lack of iron but because of a problem with the production of hemoglobin in the red blood cells.

190

What is artificial blood?

Artificial blood is a blood substitute that can be used to provide fluid volume and carry oxygen in the vessels. Two characteristics that a blood substitute should have is that it should be thinner than real blood and it should have a low affinity for oxygen so that oxygen can be delivered easily. The benefits of artificial blood are that it lessens the demand for human blood supplies and it can be given immediately without triggering a rejection in cases of massive blood loss.

Synthetic chemical compounds called perfluorocarbons are currently being studied as a substitute for red blood cells. For such a substitute to be acceptable, it needs to be: 1) able to carry oxygen and release it to tissues; 2) nontoxic; 3) storable; 4) able to function for varying periods of time in the human body; and, 5) immune-response resistant.

Individuals with thalassemia do not have enough hemoglobin or red blood cells to transport oxygen throughout the body. Mild or moderate forms of thalassemia may not require any treatment. The most common form of severe thalassemia in the United States is beta thalassemia major, also known as Cooley's anemia. Treatment usually includes frequent blood transfusions coupled with iron chelation therapy to remove the iron that accumulates in the body from the transfusions.

What is **blood doping**?

Blood doping refers to the use of artificial means to increase the number of red blood cells. Erythropoietin, a hormone produced by both the kidney and liver, increases the rate of maturation of erythrocytes in the red bone marrow. A genetically engineered form of erythropoietin (EPO) can be injected into athletes before an event, with a resulting increase of red blood cell count from 45 to 65 percent. With more red blood cells, there is more oxygen available for heavily exercising muscles. However, once an athlete becomes dehydrated, the blood can become too thick, causing clotting, stroke, or serious heart problems.

What are **leukocytes**?

Leukocytes, or white blood cells, make up less than one percent of total blood volume. Their main function is immunological defense. There are two major categories of leukocytes, which are based on their structural characteristics: granulocytes contain membrane bound granules, while agranulocytes lack membrane bound granules. The total number of white blood cells is 4,500 to 10,000 per cubic millimeter of blood. A differential white blood cell count refers to the percentage of each of the five types of white blood cells.

Leukocytes

Category	Type	Function	Normal Value
Granulocyte	Neutrophil	Chemically attracted to inflammation sites; capable of killing bacteria	50–70%
	Eosinophil	Kill parasite worms; can activate chemicals released during allergic reactions	1–3%
Agranulocyte	Basophil		0.4–1%
	Lymphocyte		25–35%
	Monocyte		4–6%

Where are **white blood cells stored**?

Most of the white blood cells in the body are stored in connective tissues or lymphatic organs. The white blood cells are released in response to areas of invasion by pathogens or injury.

How does the **white blood cell count change** with **age**?

At birth, a newborn has a higher white blood cell count (9,000 to 30,000 cells per cubic millimeter of blood), but this number falls to adult levels (4,500 to 10,000 cells per cubic millimeter of blood) within two weeks of birth. The total white blood cell count decreases slightly in the elderly.

What are the **critical values** for **white blood cell count**?

An individual with a white blood cell count of less than 500 is at high risk for infection. Leukopenia is a condition in which there are low numbers of white blood cells. A white blood cell count greater than 30,000 is an indicator of a major infection or a serious blood disorder such as leukemia. Leukocytosis is a condition in which there are very large numbers of white blood cells.

What is **diapedesis**?

Diapedesis is the ability of white blood cells to squeeze between the cells that form blood vessel walls. Once these white blood cells are outside the blood, they move through interstitial spaces using a form of primitive movement called amoeboid motion. Neutrophils and monocytes are the most active of these white blood cells. These leukocytes engulf bacterial cells, organic molecules in bacterial cells, and other large objects such as parasites. Neutrophils and monocytes frequently become so full of bacterial toxins and other related products that they also die.

What are the functions of **platelets**?

Platelets are very small 1.575×10^{-3} inches (0.004 millimeters) in diameter and 3.9×10^{-5} inches (0.001 millimeter) thick. The functions of platelets include: 1) transport of enzyme and proteins critical to clotting; 2) formation of a platelet plug to slow blood loss; and 3) contraction of a clot after it has formed, which then reduces the size of the vessel break.

Where are platelets produced?

Platelets are produced in the bone marrow as very large cells up to 0.0063 inches (0.16 millimeters) that gradually break up cytoplasm into small packets. The small packets, of which 4,000 can be released from one large cell, become the adult platelets.

What happens if **blood** does not **clot**?

If blood clots slowly or not at all, a person is at great risk for blood loss from the smallest injury. The two most common clotting disorders are hemophilia and von Willebrand disease. Hemophilia affects mostly males and is caused by a deficiency in a specific clotting factor. Von Willebrand disease is due to a deficiency in a plasma protein that interacts with specific clotting proteins. In both diseases, the severity of the condition depends on how much of the specific protein is produced.

What is an **anticoagulant**?

An anticoagulant is any substance that prevents platelets from piling up in the inner lining (endothelium) of blood vessels. Endothelial cells naturally secrete nitric oxide and prostacyclin, which prevent platelets from sticking together. Another natural anticoagulant is heparin, which is found in basophils (a type of white blood cell) and on the surface of endothelial cells. It interferes with the process of clot formation.

Platelet cells are the body's tool for repairing damage through the creation of clots. (From Rubin, E., M.D., and Farber, J.L., M.D. *Pathology.* 3rd Ed. Philadelphia: Lippincott, Williams & Wilkins, 1999.)

193

What is **plasma**?

Plasma is the liquid part of blood. It accounts for 46 to 63 percent of total blood volume. It is mostly water with a number of dissolved substances that add to its viscosity. The majority (92 percent) of the dissolved solutes are plasma proteins. Nonprotein components include metabolic waste products, nutrients, ions, and dissolved gases.

What are the major **plasma proteins**?

There are three major plasma proteins. Plasma proteins are produced by the liver, with the exception of gamma globulins, which are produced by lymphatic tissue or tissues.

Major Plasma Proteins

Type	% (by Weight)	Function
Albumin	60	Important in maintaining osmotic pressure
Globulins	36	Alpha, beta—transport proteins; gamma—antibodies released during immune response
Fibrinogen	4	Blood clot formation

What is the difference between **plasma** and **serum**?

Plasma is whole blood minus cells and serum is plasma minus clotting proteins. Serum is collected by allowing blood to clot.

How much blood does the **average-sized adult** human have?

An adult man has 5.3 to 6.4 quarts, or 1.5 gallons (5 to 6 liters), of blood, while an adult woman has 4.5 to 5.3 quarts, or 0.875 gallons (4 to 5 liters). Differences are due to the sex of the individual, body size, fluid and electrolyte concentrations, and amount of body fat.

Which **vein** is usually used to **collect blood**?

Fresh blood is usually collected from the median cubital vein (inside the elbow) in a procedure is called venipuncture. If only a small amount of blood is needed, the tip of a finger, an ear lobe, the big toe, or heel (in infants) are common areas for a pinprick.

How are components of **blood separated**?

A blood sample is centrifuged, which separates it into two components: plasma (55 percent) and formed elements (45 percent). Further centrifugation will separate plasma into proteins, water, and dissolved solutes. Additional centrifugation of the dissolved solutes will separate into platelets, leukocytes, and erythrocytes.

What are some **tests** that can be done on a collected sample of **blood**?

The table below lists common blood tests.

Who discovered the ABO system of typing blood?

The Austrian physician Karl Landsteiner (1868–1943) discovered the ABO system of blood types in 1909. Landsteiner had investigated why blood transfused from one individual was sometimes successful and other times resulted in the death of the patient. He theorized that there must be several different blood types. If a transfusion occurs between two individuals with different blood types, the red blood cells will clump together, blocking the blood vessels. Landsteiner received the Nobel Prize in Physiology or Medicine in 1930 for his discovery of human blood groups.

Blood Test	Purpose	Normal Range
Hematocrit (HCT)	% blood cells in whole blood	37–54%
Hemoglobin concentration	Hemoglobin concentration	12–18g/dl*
Red blood cell count	Number of red blood cells/microliter whole blood	4.2–6.3 million/microliter
White blood cell count	Determines total number of circulating white blood cells	600–9,000

*g/dl = grams/deciliter

Do **males or females** have a **higher hematocrit**?

Males have a higher hematocrit, since they have a greater capacity to carry oxygen in order to supply the greater muscle mass of their bodies.

What is the **amount** of **carbon dioxide** found in normal blood?

Carbon dioxide normally ranges from 19 to 50 millimeters per liter in arterial blood and 22 to 30 millimeters per liter in venous blood.

What is the **Rh factor**?

In addition to the ABO system of blood types, blood types can also be grouped by the Rhesus factor, or Rh factor, an inherited blood characteristic. Discovered independently in 1939 by Philip Levine (1900–1987) and R.E. Stetson and in 1940 by Karl Landsteiner (1868–1943) and A.S. Weiner, the Rh system classifies blood as either having the Rh factor or lacking it. Pregnant women are carefully screened for the Rh factor. If a mother is found to be Rh-negative, the father is also screened. Parents with incompatible Rh factors can have babies with potentially fatal blood problems. The condition can be treated with a series of blood transfusions.

How common is **erythroblastosis fetalis**?

Erythroblastosis fetalis, hemolytic disease of the newborn, is extremely rare today because physicians carefully track Rh status. An Rh-negative woman who might carry an Rh-positive fetus is given an injection of a drug called RhoGAM. This injection is actually composed of anti-Rh antibodies, which bind to and shield any Rh-positive fetal cells that might contact the woman's cells, sensitizing her immune system. RhoGAM must be given within 72 hours of possible contact with Rh-positive cells, including giving birth, terminating a pregnancy, miscarrying, or undergoing amniocentesis.

Which of the **major blood types** are the most **common** in the United States?

The following table lists the blood types and their rate of occurrence in the United States.

Distribution of Blood Types in the United States

Blood Type	Frequency (U.S.)
O+	37.4%
O–	6.6%
A+	35.7%
A–	6.3%
B+	8.5%
B–	1.5%
AB+	2.4%
AB–	0.6%

What are the **preferred** and permissible **blood types** for **transfusions**?

The table below lists the blood types that are best matched with other blood types.

Blood Type of Recipient	Preferred Blood Type of Donor	Permissible Blood Type(s) of Donor in an Emergency
A	A	A, O
B	B	B, O
AB	AB	AB, A, B, O
O	O	O

How **often** can **blood** be **donated**?

As long as the potential donor is in good health, whole blood may be donated every 50 days. Plasma may be donated twice a week, and platelets may be donated a maximum of 24 times per year.

What **antigens and antibodies** are associated with each **blood type**?

The table below explains the relationship between blood types and antigens and antibodies.

Blood Type	Antigen on Red Blood Cell Surface	Antibody in Plasma
A	A	Anti-B
B	B	Anti-A
AB	A and B	Neither anti-A nor anti-B
O	Neither A nor B	Both anti-A and anti-B

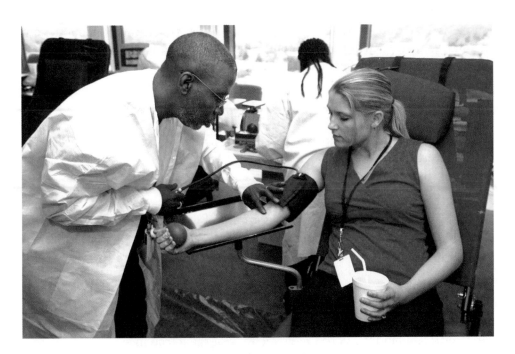

Whole blood can be safely donated every 50 days. © iStockphoto.com/David H. Lewis.

THE HEART

What is the **size and location** of the **heart**?

Heart size varies with body size. The average adult's heart is about 5.5 inches (14 centimeters) long and 3.5 inches (9 centimeters) wide, or approximately the size of one's fist. The heart is located just above the diaphragm, between the right and left lungs. One-third of the heart is located on the right size of the chest, while two-thirds is located on the left side of the chest.

How much does the **heart weigh**?

In an infant, the heart is about a thirtieth of total body weight. In an average adult, the heart is about one three-hundredth of total body weight; this equals about 11 ounces (310 grams) in males and 8 ounces (225 grams) in females.

How is **cardiac muscle** different from **skeletal muscle**?

Cardiac muscle, called the myocardium, is composed of a number of long, branching cells that are joined by intercalated discs. An intercalated disc is an area where cell membranes of adjacent cardiac muscle cells are joined. There are also small spaces in cardiac muscle cells that create a direct electrical connection between cells by allowing ions to move freely between cells. The interconnecting matrix joins cardiac muscle cells into a single, very large muscle cell called a syncytium (Latin for "joined cells"). Another difference between skeletal muscle and cardiac

197

muscle is that cardiac muscle has pacemaker cells, which initiate contractions rhythmically rather than through neural stimulation. Cardiac muscle contraction lasts about 10 times longer than skeletal muscle contractions, and cardiac muscle cannot produce sustained contractions as skeletal muscles do.

What are the **three layers** of the **heart wall**?

The wall of the heart is composed of three distinct layers: an outer epicardium, a middle myocardium, and an inner endocardium.

What is **pericarditis**?

Pericarditis is an inflammation of the pericardium, a membrane that surrounds the heart. It is frequently due to viral or bacterial infections, which produce adhesions that attach the layers of the pericardium to each other. This is a very painful condition and interferes with heart movements. Mild cases of pericarditis may resolve themselves with little treatment other than bed rest and anti-inflammatory medications. More severe cases of percarditis may require hospitalization and/or surgical removal and drainage of the fluid. If a bacterial infection is the underlying cause of pericarditis, antibiotics will be prescribed to treat the infection.

What is **angina pectoris**?

Angina pectoris (from the Latin, meaning "strangling" and "chest") is severe chest pain that occurs when the heart muscle is deprived of oxygen. It is a warning that coronary arteries are not supplying enough blood and oxygen to the heart.

How is the **heart protected** from injury?

Since the heart is continuously moving, it is protected against friction by a large pericardial sac with an outer fibrous layer and an internal serous layer. The internal layer produces fluid, which lubricates the sac in which the heart moves. The syncytium of the myocardium wraps the cavities of the heart in a continuous muscular sheet.

What are the various **chambers** of the **heart**?

The heart is divided into two upper chambers called atria (singular, atrium) and two lower chambers called ventricles. The atria are receiving chambers, where blood is delivered via large vessels, and the ventricles are pumping chambers, where blood is pumped out of the heart via large arteries.

What is **mitral valve prolapse**?

Mitral valve prolapse (MVP) is a condition in which the mitral (bicuspid) valve extends back into the left atrium when the heart beats, causing blood to leak from the atrium to the ventricle. It may be due to genetic factors or a bacterial infection, caused by *Streptococcus* bacteria. Mitral valve prolapse affects up to six percent of the U.S. population. Surgery is sometimes required to repair a valve, although most people do not require any treatment.

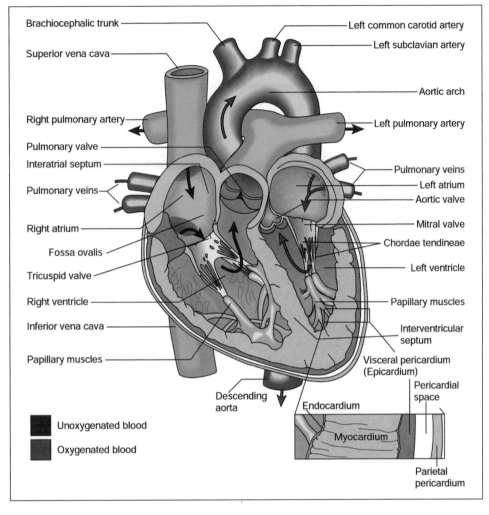

A detailed look at the heart. (From Smeltzer, S. C. O., and Bare, B. G. *Brunner and Suddarth's Textbook of Medical-Surgical Nursing.* 9th Ed. Philadelphia: Lippincott, Williams & Wilkins, 2002.)

Why is the **left ventricle larger** than the right ventricle?

Although the right and left ventricles contain equal amounts of blood, the left ventricle is larger because it has thicker walls. These thicker walls enable pressure to develop enough to push blood throughout the body. Since the right ventricle only pumps the blood to the neighboring lungs, the same ejection force is not required.

What are the **main vessels** entering and leaving the **heart**?

The main vessels entering the right side of the heart are the inferior and superior vena cava, which return low-oxygen blood to the right atrium. Blood leaves the right ventricle through the pulmonary artery to the lungs. High-oxygenated blood returns to the left atrium through the left and right pulmonary veins. All blood exits the left side of the heart through the aorta.

199

When was the **first artificial heart** implanted?

The first artificial heart was made by Dr. Robert K. Jarvik (1946–) in 1981. It was implanted in 1982 into Barney Clark. Clark lived for 112 days after the surgery.

What is **a circulatory assist device**?

A circulatory assist device, also known as a ventricular assist device, is a mechanical circulatory machine. These pumps are used on a short-term basis to allow the patient's heart to rest while it is healing. However, they have also been used on a long-term basis to support the heart of patients awaiting a heart transplant. There are three major types of devices: counterpulsation devices, cardiopulmonary assist devices, and left ventricular assist devices.

How **fast** and **how often** does the human **heart beat**?

The human heart beats 130 times per minute in infants and slows to 90 times per minute in a ten year old. By the time adulthood is reached, the heart slows to an average of 70 times per minute in men and 78 times per minute in women. The heart will beat approximately 40 million times in one year, or about 3 billion times in an average lifetime.

What are the **lubb-dupp sounds** that the heart makes?

Heart sounds are monitored by using a stethoscope. The characteristic lubb-dupp that the heart makes is due to the closing of the two sets of valves. The "lubb" is due to closing of the atrioventricular vales, and the "dupp" is due to the closing of the semilunar valves.

Who **invented** the **stethoscope**?

The stethoscope was invented in 1816 by Rene T.H. Laennec (1781–1826), a French physician. Stethoscope comes from the Greek words meaning "to study the chest."

How much blood is **pumped** by the heart?

On average, each heart contraction pumps 2.4 ounces (70 milliliters) of blood. The heart pumps 7,397 quarts (7,000 liters) of blood through the body each day.

How is the **cardiac muscle** supplied with **blood**?

Cardiac muscle has its own separate circulation, the coronary circulation. There are two large coronary arteries that supply the ventricles, with the most abundant blood supply going to the left ventricle, as this chamber has the most strenuous workload.

How is **blood flow directed** within the **heart**?

A system of valves prevents backflow both within chambers and in the large vessels exiting the heart. The atrioventricular valves are located between the right atrium and right ventricle (tricuspid valve) and left atrium and left ventricle (bicuspid valve). When the ventricles contract, blood moves back toward the atria, causing the flaps of these valves to close. The semilunar valves resemble a tripod and close after blood has exited the right ventricle (pulmonary semilunar valve) and left ventricle (aortic semilunar valve). When the ventricles are relaxed, the atrioventricular valves are open and the semilunar valves are closed. When the ventricles contract, the atrioventricular valves are closed and the semilunar valves are open.

Where is the **pacemaker** of the heart located?

The pacemaker of the heart is located in the sinoatrial (SA) node in the right atrium. Cells of the sinoatrial node generate an impulse about 75 times per minute. The pacemaker coordinates heart rate through a system of nerve fibers that spread throughout the right and left atria.

When was the **first successful pacemaker** invented?

The first successful pacemaker was developed in 1952 by Paul Zoll (1911–1999) in Boston, in collaboration with the Electrodyne Company. The device was worn externally on the patient's belt. It relied on an electrical wall socket to stimulate the patient's heart through two metal electrodes attached to the patient's chest. Wilson Greatbatch, working together with Earl Bakken, developed an internal pacemaker. It was first implanted by the surgeons Dr. William Chardack and Dr. Andrew Gage in 1960.

How can **electrical activity** of the heart be **monitored**?

The electrical activity of the heart can be monitored by an electrocardiogram. Electrodes are placed at different locations on the chest and each time the heart beats, there is a wave of electrical activity through the heart muscle. This test can detect very slight changes in the heart's electrical activity through deflections on a monitor. An electrocardiogram can be used to detect and diagnose cardiac arrhythmias, which are abnormalities in the heart's conduction system.

What are the **symptoms** of a **heart attack**?

Although some heart attacks are sudden, most heart attacks start slowly, with mild pain. The following are signs of a heart attack:

1. Chest discomfort, usually in the center of the chest and lasting more than a few minutes
2. Discomfort in other areas of the upper body, such as one or both arms, the back, neck, jaw, or stomach
3. Shortness of breath
4. Other signs such as nausea, lightheadedness, or breaking out in a cold sweat

What is **echocardiography**?

Echocardiography is a noninvasive method for studying the motion and internal vessels of the heart. This method uses ultrasound beams, which are directed into the patient's chest by a transducer. The transducer uses the ultrasonic waves, which are directed back from the heart to form an image. An echocardiogram can show internal dimensions of the chambers, valve motion, blood flow, and the presence of increased pericardial fluid, blood clots, or tumors.

How does **exercise** affect the **heart**?

Regular exercise increases the amount of blood the heart can eject with each beat, so fewer beats per minute are needed to maintain cardiac output. Exercise can increase cardiac output from 300 to 500 percent and increase heart rate up to 160 beats per minute. Individuals who exercise regularly tend to have lower resting heart rates.

BLOOD VESSELS

What are the **blood vessels** and their **function**?

Blood vessels form a closed circuit that carries blood from the heart to the organs, tissues, and cells throughout the body and then back to the heart. The blood vessels include arteries, arterioles, capillaries, venules, and veins. Arteries carry blood

---

Does your heart stop beating when you sneeze?

The heart does not stop beating when you sneeze. Sneezing, however, does affect the cardiovascular system. It causes a change in pressure inside the chest. This change in pressure affects the blood flow to the heart, which in turn affects the heart's rhythm. Therefore, a sneeze does produce a harmless delay between one heartbeat and the next, often misinterpreted as a "skipped beat."

---

away from the heart under high pressure. The arteries subdivide into smaller, thinner tubes called arterioles. As the arterioles approach capillaries, the walls of the vessels become very thin. Capillaries have the smallest diameter of all the blood vessels. They connect the arterioles with the venules. Venules continue from the capillaries to form the veins.

What are the differences between **arteries** and **veins**?

Both arteries and veins have three tissue layers: the inner lining (endothelium), the middle layer (smooth muscle), and the outer layer (connective tissue). However, the walls of the arteries are much stronger and thicker to accommodate the blood under high pressure as it exits the heart. Many veins have valves to help return the blood to the heart.

What is an **aneurysm**?

An aneurysm is a bulge in the weakened wall of an artery, most often the aorta. It is similar to what one would see when there is a bubble in the wall of a garden hose. If an aneurysm becomes large enough, it can burst, resulting in a stroke if it is in a brain artery or massive hemorrhage if it is the wall of the aorta. If an aneurysm bursts, the massive bleeding is often fatal.

What is the difference between **arteriosclerosis** and **atherosclerosis**?

Arteriosclerosis, also known as hardening of the arteries, occurs when the arterial walls thicken and then harden as calcium deposits form. If the coronary vessels are affected, this is known as coronary artery disease. Atherosclerosis is another type of hardening of the arteries in which lipids, particularly cholesterol, build up on the side arterial walls. Risk factors for atherosclerosis include cigarette smoking, a high fat/high cholesterol diet, and hypertension.

What is the **largest artery** in the human body?

The aorta is the largest artery in the human body. In adults, it is approximately the size of a garden hose. Its internal diameter is 1 inch (2.5 centimeters) and its wall is about 0.079 inches (0.2 centimeters) thick.

What is the **largest vein** in the human body?

The largest vein in the human body is the inferior vena cava, the vein that returns blood from the lower half of the body back to the heart.

What is a **capillary bed**?

A capillary bed is a spiderweb-like network that connects the arterial system with the venous system in a particular body region. A capillary bed may receive blood from more than one artery.

What are the **functions of capillaries**?

Capillaries are perhaps the most important of the blood vessels because they are the primary exchange points of the cardiovascular system. Gases, nutrients, and metabolic byproducts are exchanged between the blood in capillaries and the tissue fluid surrounding body cells. The materials exchanged move through capillary walls by diffusion, filtration, and osmosis.

How **big** are **capillaries**?

The diameter of a capillary is about 0.0003 inches (0.0076 millimeters), which is just about the same as a single red blood cell. A capillary is only about 0.04 inches long (1 millimeter). If all the capillaries in a human body were placed end to end, the collective length would be approximately 25,000 miles (46,325 kilometers), which is slightly more than the circumference of the earth at the equator: 24,900 miles (46,139 kilometers).

What is the **collective length** of all **blood vessels** in the human body, including arteries, arterioles, capillaries, venules, and veins?

The length of all the blood vessels in the circulatory system is approximately 60,000 miles (96,500 kilometers). If laid end to end, the body's blood vessels would encircle the earth more than two times.

Where is **most of the blood** in the body at any one time?

Blood volume is not evenly distributed among the different types of vessels. Due to the expandable properties of veins, a vein will stretch about eight times more than an artery of corresponding size. At rest, the venous system thus contains about 65 to 70 percent of total blood volume, with the heart, arteries, and capillaries containing 30 to 35 percent of total blood volume. About one-third of the blood in the venous system is found in the liver, bone marrow, and skin.

What are **varicose veins**?

Varicose veins are distended veins, usually in the superficial veins of the thighs and legs. Varicose veins are caused by the valves inside the veins becoming stretched so

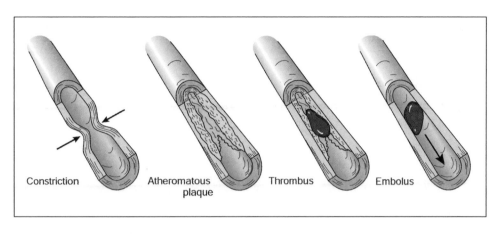

Constriction   Atheromatous   Thrombus   Embolus
                       plaque

Constriction, the build-up of plaque, or clots can restrict blood flow in veins or arteries. (From Willis, M.C. *Medical Terminology: A Programmed Learning Approach to the Language of Health Care*. Baltimore: Lippincott, Williams & Wilkins, 2002.)

they no longer close completely. The affected veins then become filled with blood. There are usually no serious medical problems associated with varicose vein.

Why does **blood** in the **veins look blue**?

Since venous blood is oxygen-poor blood, it is not as bright red as arterial blood. It appears as a deep, dark-red, almost purplish color. Seeing "blue blood" in veins through the skin is a combination of light passing through the skin and the oxygen-poor blood.

How does **arterial bleeding** differ from **venous bleeding**?

Each type of vessel (artery, vein, capillary) has distinct bleeding characteristics. Arterial bleeding is characterized by spurts of bright-red blood; spurting each time the heart beats. Arterial bleeding is serious and difficult to control. Venous bleeding occurs in a steady flow, and the blood is dark-red, almost maroon, in color. Since capillaries are so small, capillary bleeding is a slow, oozing flow that carries with it a higher risk of infection than either arterial or venous bleeding.

CIRCULATION

What is **pulse**?

Pulse is the alternate expansion and recoil of an artery, which can be felt in an artery close to the body's surface because of the rhythmic ejection of blood from the heart into the aorta, which causes an increase and decrease of pressure in the vessel. Pulse provides important information about the heart action, blood vessels, and circulation. A fast pulse rate may indicate the presence of an infarction or dehydration. In a medical emergency, the pressure of a pulse will help determine if a person's heart is pumping.

Where is the pulse most easily found?

Pulse can be found anywhere an artery is near the surface or over a bone. The body sites where pulse is most easily felt are:

- Wrist (radial artery)
- Temporal artery (in front of ear)
- Common carotid artery (along the lower margin of the jaw)
- Facial artery (lower margin of the lower jawbone)
- Brachial artery (at the bend of elbow)
- Popliteal artery (behind the knee)
- Posterior tibial artery (behind the ankle)
- Dorsalis pedis artery (on the upper surface of the foot).

What are the normal values for the human pulse rate?

The following are normal values:

Age	Pulse Rate
Newborn	100–160 beats/minute
1–10 years old	70–120 beats/minute
10 to adulthood	60–100 beats/minute
Athletes	40–60 beats/minute

How fast does blood move in vessels?

Blood flow refers to the volume of blood flowing through a vessel or group of vessels during a specific time. It is measured in milliliters per minute (mL/min). Blood flow during rest averages three to four mL/minute per 100 grams of muscle tissue, but may increase to 80 mL/minute or more during exercise. The distance blood flows during a specific time period is its velocity. Blood velocity is measured as centimeters/second (cm/s). In general, blood velocity is greatest in larger vessels and decreases in vessels with a smaller diameter. The velocity of blood in the aorta is approximately 30 cm/s, in arterioles 1.5 cm/s, in capillaries 0.04 cm/s, in venules 0.5 cm/s, and in the venae cavae 8 cm/s.

How do veins return blood to the heart?

Veins in the chest rely on the processes of inspiration and expiration of the lungs and the subsequent movement of the diaphragm to "pump" blood back to the heart. Other "pump" mechanisms are contractions of the skeletal muscles, which squeeze the veins within the muscles throughout the body. However, it is the semilunar valves found in veins that prevent blood backflow when the skeletal muscles relax.

What are the sounds of Korotkoff?

The sounds of Korotkoff are produced when blood pressure is taken. They are named after Dr. Nikolai Korotkoff (1874–1920), a Russian physician who first described them in 1905, when he used a stethoscope to listen to the sounds of blood flowing through an artery.

What do the **readings** of **blood pressure** mean?

Blood pressure is monitored using a sphygmomanometer (*sphygmos,* meaning "pulse," and *manometer,* meaning a device for measuring pressure). Blood pressure is the pressure in the arterial system in the largest vessels near the heart as the heart pushes blood through vessels. It is measured in millimeters of mercury. The two numbers for blood pressure reflect two different pressures: systolic pressure and diastolic pressure. Systolic pressure (the upper number in a blood pressure reading) is the pressure of the blood against the arterial walls when the ventricles are contracting. Diastolic pressure (the lower number in a blood pressure reading) is the lowest point at which sounds can be heard as the blood flows through the smaller arteries. A blood pressure reading below 120/80 mm Hg is considered normal.

What is **hypertension**?

Hypertension is a sustained condition when the blood pressure exceeds 140/90 mm Hg. It is estimated that about 30 percent of people ages 50 and over have hypertension.

What is **hypotension**?

Hypotension, or low blood pressure, is a systolic pressure below 100mm Hg. Most commonly, it is due to overly aggressive treatment for hypertension.

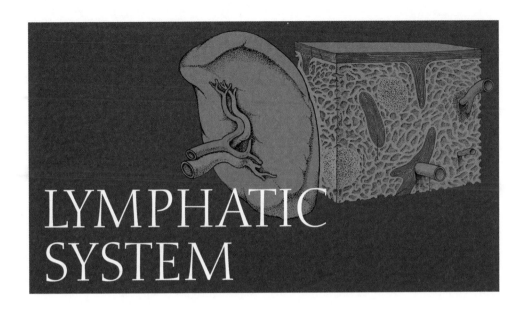

LYMPHATIC SYSTEM

INTRODUCTION

Which **branch of medicine** deals with the **responses of the body** to **foreign substances**?

Immunology (from the Latin *immunis,* meaning "free from services or obligations," and the Greek *ology,* meaning "the study of") is the study of cells and tissues that carry out immune responses.

What are the primary **functions** of the **lymphatic system**?

The lymphatic system is responsible for maintaining proper fluid balance in tissues and blood, in addition to its role defending the body against disease-causing agents. The primary functions of the lymphatic system are: 1) to collect the interstitial fluid that consists of excess water and proteins and return it to the blood; 2) to transport lipids and other nutrients that are unable to enter the bloodstream directly; and 3) to protect the body from foreign cells and microorganisms.

What are the three major **components** of the **lymphatic system**?

The lymphatic system consists of the lymphatic vessels, lymph, and lymphoid organs. Together these components form a network that collects and drains most of the fluid that seeps from the bloodstream and accumulates in the space between cells.

What are the **primary cells** of the **lymphatic system**?

The primary cells of the lymphatic system are lymphocytes. There are three types of lymphocytes: T cells, B cells, and NK cells. T cells account for approximately 80 percent of the circulating lymphocytes. They are thymus-dependent and are the primary cells that provide cellular immunity. B cells, which are derived from the bone marrow, account for 10 to 15 percent of the circulating lymphocytes. They are responsible for antibody-mediated immunity. NK (natural killer) cells account for the remaining 5 to 10 percent of the circulating lymphocytes. They attack foreign cells, normal cells infected with viruses, and cancer cells that appear in normal tissues.

Which **other cells** play a role in the **immune system**?

White blood cells, including macrophages, neutrophils, eosinophils, basophils, and mast cells, all have active roles in the immune system. Macrophages, in particular, are important phagocytic cells that destroy many pathogens.

What are **mast cells**?

Mast cells are specialized cells of connective tissue. They release heparin, histamine, leukotrienes, and prostaglandins to stimulate the inflammatory response.

How many lymphocytes are in the body?

There are approximately 10 trillion lymphocytes, weighing over 2.2 pounds (1 kilogram) in the body. They account for 20 to 30 percent of the circulating white blood cell population.

How **long** do **lymphocytes survive**?

Lymphocytes have relatively long life spans. Nearly 80 percent survive for four years and some last as much as 20 years or more. They spend varying amounts of time in the blood, lymphatic vessels, and lymphatic organs. New lymphocytes are produced in the bone marrow and lymphoid tissues.

What is the **composition** of **lymph**?

Lymph is similar in composition to blood plasma. The main chemical difference is that lymph does not contain erythrocytes. It also contains a much lower concentration of protein than plasma since most protein molecules are too large to filter through the capillary wall. Lymph contains water, some plasma proteins, elec-

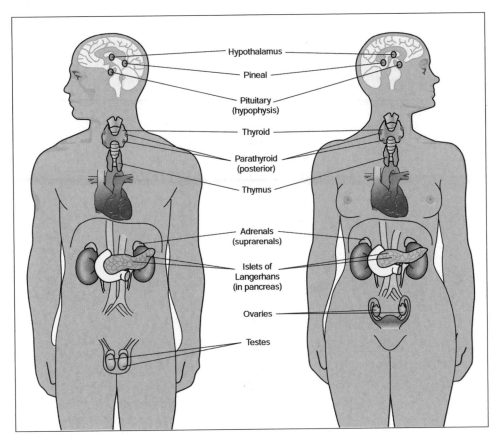

The lymphoid organs. (From Smeltzer, S. C., and Bare, B. G. *Textbook of Medical-Surgical Nursing.* 9th Ed. Philadelphia: Lippincott, Williams & Wilkins, 2000.)

trolytes, lipids, leukocytes, coagulation factors, antibodies, enzymes, sugars, urea, and amino acids.

How much lymph is in the body?

The body contains approximately 1 to 2 quarts (1 or 2 liters) of lymph, which accounts for one to three percent of body weight.

How much **fluid** does the **lymphatic system return** to the **circulatory system**?

The lymphatic system returns slightly more than 3 quarts (3 liters) of fluid from the tissues to the circulatory system on a daily basis.

What is an **autoimmune disease**?

An autoimmune disease is one in which the body triggers an immune response against the body's own cells and tissues. The cause of most autoimmune diseases is unknown. Autoimmune diseases can affect almost every organ and body system.

They may be systematic (affecting and damaging many organs) or localized (affecting only a single organ or tissue).

Autoimmune Diseases of Various Body Systems

Body System	Autoimmune Diseases
Blood and blood vessels	Autoimmune hemolytic anemia; pernicious anemia; polyarteritis nodosa; systemic lupus erythematosus; Wegener's granulomatosis
Digestive tract (including the mouth)	Autoimmune hepatitis; Behçet's disease; Crohn's disease; primary bilary cirrhosis; scleroderma; ulcerative colitis
Eyes	Sjögren's syndrome; Type 1 diabetes mellitus; uveitis
Glands	Graves' disease; thyroiditis; Type 1 diabetes mellitus
Heart	Myocarditis; rheumatic fever; scleroderma; systemic lupus erythematosus
Joints	Ankylosing spondylitis; rheumatoid arthritis; systemic lupus erythematosus
Kidneys	Glomerulonephritis; systemic lupus erythematosus; Type 1 diabetes mellitus
Lungs	Rheumatoid arthritis; sarcoidosis; scleroderma; systemic lupus erythematosus
Muscles	Dermatomyositis; myasthenia gravis; polymyositis
Nerves and brain	Guillian-Barré syndrome; multiple sclerosis; systemic lupus erythematosus
Skin	Alopecia areata; pemphigus/pemphigoid; psoriasis; scleroderma; systemic lupus erythematosus; vitiligo

How does the **immune system fail** in immunodeficiency **diseases?**

The immune system may fail in one of two ways in immunodeficiency diseases. Either the immune system fails to develop normally, as in severe combined immunodeficiency disease (SCID), or the immune response is blocked in some way, as in acquired immunodeficiency syndrome (AIDS). Treatment with immunosuppressive agents, such as radiation or specific drugs, may also result in immunodeficiency diseases.

LYMPHATIC VESSELS AND ORGANS

What are the different **lymph vessels**?

The smallest lymph vessels are lymphatic capillaries, which originate in the peripheral tissues. They are larger in diameter than blood capillaries, but have a thinner wall. The lymphatic capillaries have a unique structure that allows interstitial fluid to flow into them, but not out. Lymph flows from the lymphatic capillaries into larger lymph vessels that lead toward the trunk of the body. The lymphatics continue to join together, finally forming two large ducts: the right lymphatic duct and the thoracic duct.

Do all **tissues** have **lymphatic capillaries**?

Lymphatic capillaries are found in almost every tissue and organ of the body. They are not found in avascular tissues (tissues that lack a blood supply), such as cartilage, the epidermis and cornea of the eye, the central nervous system, portions of the spleen, and red bone marrow.

Which vessels are the **large collecting ducts**?

The right lymphatic duct and thoracic duct are the large collecting ducts. The right lymphatic duct drains the right side of the head, right upper limb, right thorax and lung, right side of the heart, and the upper portion of the liver. The contents of the right lymphatic duct drain into the right subclavian vein and are returned to the blood. The thoracic duct, the largest collecting duct, receives lymph from the three-quarters of the body including the left side of the head, neck, chest, the left upper limb, and the entire body below the ribs. It drains into the left subclavian vein.

What is the **route** the **fluid travels** in the body?

Blood flows from the arteries to the capillaries, with a portion leaking into the interstitial spaces. Once the fluid leaves the interstitial spaces and enters the lymphatic capillaries, it is called lymph. It flows from the lymphatic capillaries through the lymphatic vessels and lymph nodes to the lymphatic ducts, eventually entering either the right lymphatic duct or the thoracic duct. It finally drains into the subclavian veins and is returned to the blood.

How do **lymphoid tissues** differ from **lymphoid organs**?

Lymphoid tissues are connective tissues that contain large numbers of lymphocytes. Lymphoid organs are surrounded by a fibrous capsule that separates them from surrounding tissues.

213

Where are the most **lymphoid nodules located** in the body?

Lymphoid nodules are found in the connective tissues lining the digestive, urinary, reproductive, and respiratory tracts. They are small, oval-shaped, and approximately a millimeter in diameter. They are not surrounded by a fibrous capsule. The collection of lymphoid tissues lining the digestive system is called mucosa-associated lymphoid tissue (MALT) because they are found in the mucous membranes lining the digestive tract. Clusters of lymphoid tissue found in the intestine and appendix are called aggregated lymph nodules, or Peyer's patches. Tonsils are a group of lymphoid tissues found at the junction of the oral cavity, nasal cavity, and throat.

What are the **cancers** of the **lymphatic system**?

Cancers that start in the lymphoid tissue are called lymphomas (from the Latin *lympha*, meaning "water," as in the watery appearance of lymph, and from the Greek *oma*, meaning "tumor"). The two main types of lymphomas are Hodgkin's lymphoma (also called Hodgkin's disease) and non-Hodgkin's lymphoma. There are several different subtypes of both Hodgkin's lymphoma and non-Hodgkin's lymphoma.

Where are the **most common sites** in the body for **Hodgkin's disease** to begin?

Since lymphatic tissue is found throughout the body, Hodgkin's disease can start almost anywhere in the body. The most common sites for Hodgkin's disease to begin are the lymph nodes in the chest, in the neck, or under the arms. It travels through the lymph from one lymph node to another.

Who first **described Hodgkin**'s disease?

Hodgkin's disease was first described by the British physician Dr. Thomas Hodgkin (1798–1866) in 1832 in his paper "On Some Morbid Appearances of the Absorbent Glands and Spleen," published in London's *Medico-Chirurgical Trans-*

actions. It was not until 1865 that another British physician, Dr. Samuel Wilks (1824–1911), named the medical condition Hodgkin's disease. Although Wilks described the same disease independently, he became familiar with the earlier work of Hodgkin and named the disease after him in a paper entitled "Cases of Enlargement of the Lymphatic Glands and Spleen, (or, Hodgkin's Disease) with Remarks."

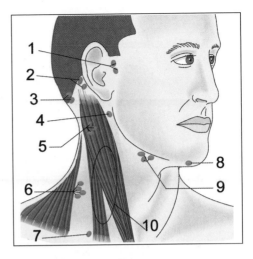

Many lymph nodes are located in the neck region: 1 = Preauricular; 2 = Occipital; 3 = Posterior auricular; 4 = Tonsillar; 5 = Superficial cervical; 6 = Posterior cervical; 7 = Supraclavicular; 8 = Submental; 9 = Submandibular; and 10 = Deep cervical chain. (From Bickley, L.S., and Szilagyi, P. *Bate's Guide to Physical Examination and History Taking.* 8th Ed. Philadelphia: Lippincott, Williams & Wilkins, 2003.)

What is the **role** of the **lymphatic system** in **metastatic cancer**?

The lymphatic vessels are often a means for transporting metastasizing cancer cells. Cancer cells often enter the lymph nodes and establish secondary cancers there. Examination and analysis of the lymph nodes provides valuable information on the spread of the cancer and helps to determine the appropriate therapy.

How many tonsils does the average person have?

Most people have five tonsils. These include a single pharyngeal tonsil, often referred to as the adenoid, located in the posterior wall of the upper part of the throat. A pair of palatine tonsils is found at the back of the mouth; a pair of lingual tonsils is located at the base of the tongue.

Which tonsils are most commonly removed during a **tonsillectomy**?

The palatine tonsils are the ones most commonly removed during a tonsillectomy. Normally, tonsils function to prevent infection. However, when the tonsils are frequently infected, physicians may recommend removal of the infected tissues when they do not respond to noninvasive treatment, such as antibiotics.

What is **appendicitis**?

The appendix is a small, blunt-ended tube composed of lymphatic tissue. It is considered part of the large intestine, although the tissues are different from the tissues of the large intestine. Appendicitis is an infection of the appendix that begins in the lymphoid nodules. Treatment often includes removal of the appendix, either with traditional surgery or laproscopically.

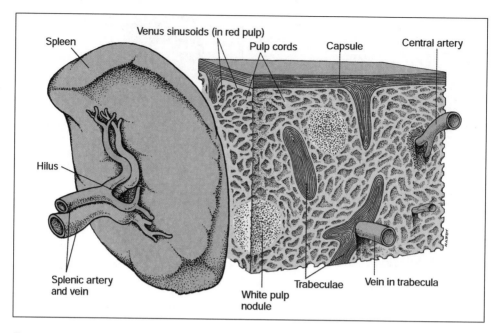

The spleen, with detailed cross section at right. (From *Stedman's Medical Dictionary*. 27th Ed. Baltimore: Lippincott, Williams & Wilkins, 2000.)

What are the major **lymphoid organs**?

The major lymphoid organs are the lymph nodes, thymus, and spleen. There are approximately 600 lymph nodes, ranging in diameter up to 1 inch (about 25 millimeters), scattered along the lymphatic vessels. The greatest concentrations of lymph nodes are found at the neck, armpit, thorax, abdomen, and groin.

The thymus serves both as an endocrine organ that secretes hormones and as a lymphoid organ. It consists of a large number of lymphocytes, many of which become specialized T cells. The spleen is the largest lymphoid organ. It is about 5 inches (12 centimeters) long and weighs 5.6 ounces (160 grams).

What are the **functions** of the **spleen**?

The primary function of the spleen is the filtering of blood and removal of abnormal blood cells by phagocytosis. The spleen also stores iron from worn-out blood cells, which is then returned to the circulation and used by the bone marrow to produce new blood cells. The immune reaction begins in the spleen with the activation of immune response by B cells and T cells in response to antigens in the blood.

How serious is **damage** to the **spleen**?

Damage to the spleen, which is often the result of an injury caused by a blow to the left side of the abdomen, may be life-threatening. Since the spleen is a fragile organ, an injury can easily rupture it, resulting in serious internal bleeding, hemorrhaging, circulatory shock, and even death. Once the spleen ruptures, the only remedy is to surgically remove it in a procedure called a splenectomy.

What are **swollen "glands"**?

The condition commonly referred to as "swollen glands" is really enlarged lymph nodes. The lymph nodes were originally referred to as lymph glands (from the Latin *glans,* meaning "acorn") because they resembled acorns. Unlike true glands, the lymph nodes do not secrete fluids, so they are now called lymph nodes (from the Latin *node,* meaning "knob"). The term "swollen glands" has been retained to describe the condition of a slight enlargement of the lymph nodes along the lymphatic vessels draining a specific region of the body. It generally indicates an inflammation or infection of peripheral structures.

NONSPECIFIC DEFENSES

What are **nonspecific defenses**?

Nonspecific defenses do not differentiate between various invaders. Barriers such as skin and the mucous membrane lining the respiratory and digestive tracts; phagocytic white blood cells; inflammation; fever; and chemicals are nonspecific defenses. The nonspecific defenses are the first to respond to a foreign substance in the body.

How does the **skin** act as a **barrier** to protect the body?

The tightly packed cells of the dermis and epidermis become a barrier that does not permit pathogens to enter the body. The acidity of the skin (pH 3 to 5) and the sebum create an environment that does not welcome microorganisms. Perspiration helps to wash away microbes from the skin. Similarly, tears wash away foreign bodies from the eyes.

How do the **mucous membranes** provide a **defense** against microorganisms?

Mucous membranes line the epithelial layer of many of the body cavities, such as the nasal passages, respiratory tract, and digestive tract. Since the mucous is viscous and slightly sticky, it traps many microorganisms, preventing them from attaching to the epithelium or entering the tissue. Some of these membranes are also ciliated, which helps to move trapped particles away from the body.

What are the four **symptoms** of **inflammation**?

The symptoms of inflammation, or the inflammatory response, are redness, heat, swelling, and pain. Inflammation is a response to tissue damage caused by injury, irritants, pathogens, distortion or disturbance of cells, and extreme temperatures. There are three stages to the inflammatory response: 1) vasodilation, allowing more blood to flow to the damaged tissue, and increased permeability of the blood vessels; 2) migration of phagocytes to the site of tissue damage; and 3) repair.

What are **natural killer (NK) cells**?

Natural killer (NK) cells are specialized lymphocytes that are able to recognize abnormal cells and destroy them. NK cells are also able to destroy certain tumor cells.

Which **chemical substances** act as **nonspecific defenses**?

Interferons and complement are nonspecific antiviral and antibacterial substances. Interferons are small proteins produced by virus-infected cells. They are an important defense against infection by many different viruses. Complement is a group of 11 special proteins found in blood plasma. These proteins supplement or enhance ("complement") certain immune, allergic, and inflammatory reactions.

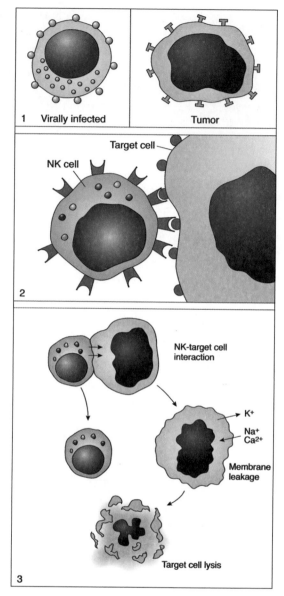

Natural killer (NK) cells target and destroy abnormal or invading cells in the body. (From Rubin, E., M.D., and Farber, J. L., M.D. *Pathology*. 3rd Ed. Philadelphia: Lippincott, Williams & Wilkins, 1999.)

What are the **two pathways** of **complement activation**?

The two routes for complement activation are the classical pathway and the alternative pathway. The classical pathway is activated by antigen-antibody interaction.

The alternative pathway is activated as a reaction to foreign bodies in the absence of antigens.

What is the **role** of **fever** in infection?

Normal body temperature is 98.6°F (37.2°C). Fever is defined as a higher-than-normal body temperature. Certain pathogens and bacterial toxins may stimulate the release of pyrogens (proteins that regulate body temperature) such as interleukin-1. Increases in body temperature increase the metabolic rate and may speed up body reactions that aid to resolve infections. Fever may also inhibit the growth of certain microbes. Fever appears to stimulate the liver to hoard substances that bacteria require, helping to decrease bacterial growth.

SPECIFIC DEFENSES

What is **specific resistance**?

Specific resistance, or immunity, is the production of specific types of cells or specific antibodies to destroy a particular antigen.

What is an **antigen**?

An antigen is a substance that triggers the immune response, causing the body to form and produce specific antibodies.

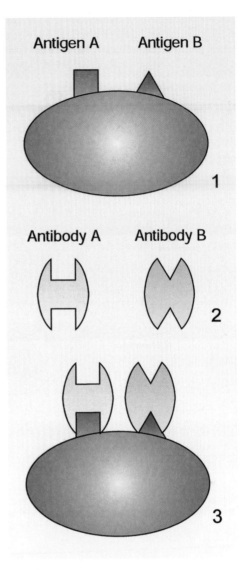

How antigens work: 1) The invading organism with antigen A and antigen B on its cell surface; 2) Antibodies A and B are produced in response to antigens A and B, respectively; and 3) The specific antibodies bind with the specific corresponding antigens, rendering them harmless. (From Smeltzer, S.C., and Bare, B.G. *Textbook of Medical-Surgical Nursing*. 9th Ed. Philadelphia: Lippincott, Williams & Wilkins, 2000.)

What is an **antibody**?

An antibody is a protein produced by B cells in response to an antigen. Antibodies are able to neutralize the antigens that provoke their production.

How many **classes** of **antibodies** have been identified?

There are five classes of antibodies, known as immunoglobulins (Igs).

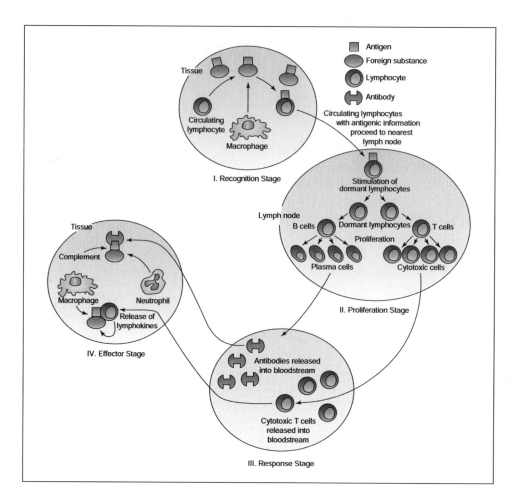

Antibodies, antigens, and lymphocytes work together to fend off disease. (From Smeltzer, S. C., and Bare, B. G. *Textbook of Medical-Surgical Nursing.* 9th Ed. Philadelphia: Lippincott, Williams & Wilkins, 2000.)

Classes of Immunoglobulins

Class	Description
IgG	Accounts for 80% of all antibodies in the blood; found in blood, lymph, and the intestines; the only antibody that crosses the placenta from mother to fetus; provides resistance against many viruses, bacteria, and bacterial toxins
IgA	Accounts for 10% to 15% of all antibodies in the blood; found mostly in secretions such as sweat, tears, saliva, and mucus; attack pathogens before they enter internal tissues; levels reduce under stress lowering resistance
IgM	Accounts for 5% to 10% of all antibodies in the blood; found in blood and lymph; first antibody secreted after exposure to an antigen; includes the anti-A and anti-B antibodies of ABO blood, which bind to A and B antigens during incompatible blood transfusions
IgD	Accounts for 0.2% of all antibodies in the blood; found in blood, lymph, and the surfaces of B cells; plays a role in the activation of B cells

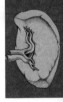

Class	Description
IgE	Accounts for less than 0.1% of all antibodies in the blood; found on the surfaces of mast cells and basophils; stimulates cells to release histamine and other chemicals that accelerate inflammation; plays a role in allergic reactions

How do **T cells** differ from **B cells**?

T cells and B cells are both lymphocytes, which respond to the presence of specific antigens. T cells arise from cells that originate in the bone marrow and then migrate to the thymus gland, where they mature. They are responsible for cell-mediated immunity and attack specific foreign cells in the body. B cells develop in the red bone marrow. They are responsible for producing and secreting specific antibodies.

Do all **B cells** produce the **same antibodies**?

No, each B cell is programmed to produce one specific antibody. For example, one B cell produces the antibody that blocks a virus that is responsible for the common cold, while a different B cell produces the antibody that attacks the bacterium that causes pneumonia.

What are **plasma cells**?

Plasma cells are large, antibody-producing cells that are derived from B cells. Each plasma cell that is derived from a single B cell manufactures millions of identical antibodies to fight the antigen (virus, microbe, or other foreign tissue/substance).

What are the **two forms** of **immunity**?

Immunity may be either innate or acquired. Innate immunity is present at birth and has no basis to prior exposure to the antigen involved. Certain innate immunity is species-specific. For example, humans are not susceptible to diseases of cats and dogs. Additionally, certain individuals are not susceptible to some diseases that other individuals are susceptible to. Acquired immunity is not present at birth, but rather is acquired following exposure to a particular antigen.

How does **active immunity** differ from **passive immunity**?

Active immunity develops when the body manufactures its own antibodies following exposure to an antigen. Active immunity may occur naturally as a result of exposure to foreign antigens or artificially via vaccinations.

Passive immunity is produced by transferring antibodies from one individual to another. One example of passive immunity is when maternal antibodies cross the placenta and provide protection to the fetus. Maternal antibodies are also passed from mother to infant during breastfeeding. Artificial passive immunity occurs when antibodies are injected into the body. Passive immunity is used to treat rabies, tetanus, and rattlesnake bites.

What are **monoclonal antibodies**?

Monoclonal antibodies are identical antibodies produced in a laboratory from a single B cell. Researchers make monoclonal antibodies by injecting a mouse with a target antigen and then fusing B cells from the mouse with another long-lived cell. The resulting hybrid cell becomes a type of antibody factory, turning out identical copies of antibody molecules specific for the target antigen. Monoclonal antibodies are used to treat cancer, some viral infections, inflammatory diseases, some cardiovascular diseases, and organ transplant rejections.

When was the **term "antibiotic"** first used?

Antibiotics are chemical products or derivatives of certain organisms that inhibit the growth of or destroy other organisms. The term "antibiotic" (from the Greek *anti,* meaning "against," and *biosis,* meaning "life") refers to its purpose in destroying a life form. In 1889 Paul Vuillemin (1861–1932) used the term "antibiosis" to describe bacterial antagonism. He had isolated pyocyanin, which inhibited the growth of bacteria in test tubes but was too lethal to be used in disease therapy. In was not until the mid-1940s that Selman Waksman (1888–1973) used the term "antibiotic" to describe a compound that had therapeutic effects against disease. Waksman received the Nobel Prize in Physiology or Medicine in 1952 for his discovery of streptomycin. Streptomycin was the first antibiotic effective against tuberculosis.

How do **antibiotics destroy** an **infection**?

Antibiotics function by weakening the cell wall, or interfering with the protein synthesis or RNA synthesis of the bacterial cell. For example, penicillin weakens the cell wall to the point that the internal pressure causes the cell to swell and eventually burst. Various antibiotics are more effective against different bacteria.

Why is it **difficult** to treat **viral infections** with medications?

Antibiotics are ineffective against viral infections because viruses lack the structures (e.g., a cell wall) with which antibiotics can interfere. In general, it is difficult to treat viral infections with medications without affecting the host cell, as viruses use the host cell's machinery during replication. Several antiviral drugs have been developed that are effective against certain viruses.

When was the earliest documented case of human immunodeficiency virus (HIV)?

The earliest known case of HIV-1 in a human was from a blood sample collected in 1959 from a man in Kinshasa, Democratic Republic of Congo. The virus has existed in the United States since the mid-1970s. Between 1979 and 1981 physicians in New York City and Los Angeles reported treating a number of male patients who had engaged in sex with other men for rare types of pneumonia, cancer, and other illnesses not usually found in individuals with healthy immune systems.

Some Antiviral Drugs Used against Diseases

Disease	Viral Pathogen	Antiviral Drug
AIDS	Human immunodeficiency virus	Azidathymidine (AZT), didanosine, dideoxycytosine
Chronic hepatitis	Hepatitis B or C	α-interferon
Genital herpes, shingles, chickenpox	Herpes virus	Acyclovir, idoxuridine, trifluridine, vidarabine
Influenza A	Influenza	Amatadine

What **naturally occurring substance** provides protection against **viral infections**?

Interferons protect the adjacent cells against viral penetration. Interferons are glycoproteins produced by body cells upon exposure to a virus. In 1957 Alick Isaacs (1921–1967) and Jean Lindenmann (1924–) identified a group of over 20 substances that were later designated as alpha, beta, and gamma interferons.

When was the **term "acquired immunodeficiency syndrome"** first used?

Public health officials in the United States began to use the term "acquired immunodeficiency syndrome" (AIDS) in 1982 to describe the occurrences of opportunistic infections, Kaposi's sarcoma (a kind of cancer), and *Pneumocystis carinii* pneumonia in previously healthy people.

What are the **symptoms** and signs of **HIV**?

The warning signs of HIV infection include night sweats, prolonged fevers; severe weight loss; persistent diarrhea; red, brown, pink, or purplish blotches on or under the skin or inside the mouth, nose, or eyelids; persistent dry cough; swollen lymph nodes in the armpits, groin, or neck; white spots or unusual blemishes on the tongue, in the mouth, or in the throat; memory loss, depression, and other neurological disorders. However, the only way to determine if you are infected is to be tested for HIV.

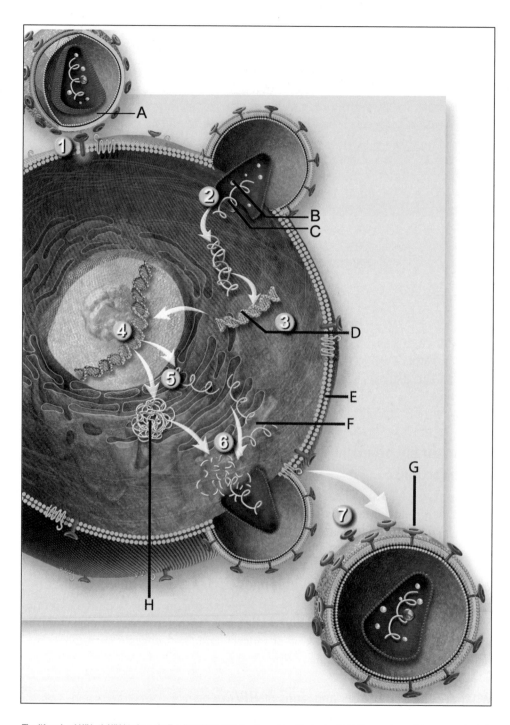

The lifecycle of HIV: 1) HIV binds to the T-cell; 2) Viral RNA is released ino the host cell; 3) Reverse transcriptase converts viral RNA into viral DNA; 4) Viral DNA enters the T-cell's nucleus and inserts itself into the T-cell's DNA; 5) The T-cell begins to make copies of the HIV components; 6) Protease (an enzyme) helps create new virus particles; 7) The new HIV virion (virus particle) is released from the T-cell; A=HIV virion (virus particle); B=Viral RNA; C=Reverse transcriptase; D=Viral DNA; E=T-cell; F=Viral RNA; G=New HIV virion; H=HIV proteins. *Anatomical Chart Co*.

How is **HIV diagnosed**?

The only accurate way to diagnose HIV is through antibody testing. The immune system of an individual infected with HIV will begin to produce HIV antibodies to fight the infection. Although these antibodies are ineffective in destroying the HIV virus, their presence is a positive indication of the presence of HIV.

What is the **difference** between human immunodeficiency virus (**HIV**) and **AIDS**?

The term AIDS applies to the most advanced stages of HIV infection. The Center for Disease Control's (CDC) definition of AIDS includes all HIV-infected people who have fewer than 200 CD4+ T cells per cubic millimeter of blood. (Healthy adults usually have CD4+ T cell counts of 1,000 or more.) The definition also includes 26 clinical conditions (mostly opportunistic infections) that affect people with advanced HIV disease.

What are some **complications of AIDS**?

The immune system of individuals with AIDS is severely compromised and weak, so the body cannot fight off certain bacteria, viruses, fungi, parasites, and other microbes. Opportunistic infections can easily establish themselves. In addition, people with AIDS are prone to developing various cancers, especially those caused by viruses such as Kaposi's sarcoma and cervical cancer, or cancers of the immune system known as lymphomas. These cancers are usually more aggressive and difficult to treat in people with AIDS.

How many **individuals** are **infected** with **HIV/AIDS**?

According to the World Health Organization (WHO), there are 40.3 million people worldwide living with HIV. An estimated 4.9 million people (adults and children) were newly infected with HIV in 2005.

Worldwide Distribution of HIV/AIDS (2005)

Region	People Infected	Newly Diagnosed Cases
North America	1.2 million	43,000
Caribbean	300,000	30,000
Latin America	1.8 million	200,000
Western & Central Europe	720,000	22,000
North Africa & Middle East	510,000	67,000
Sub-Saharan Africa	25.8 million	3.2 million
Eastern Europe & Central Asia	1.6 million	270,000
East Asia	870,000	140,000
South & Southeast Asia	7.4 million	990,000
Oceania	74,000	8,000
TOTAL	40.3 million	4.9 million

225

How is the **HIV virus transmitted** from one person to another?

HIV is transmitted via unprotected sexual contact with an infected partner or through contact with infected blood. Rigorous screening of the blood supply and heat-treating techniques for donated blood have reduced the rate of transmission via blood transfusions to a very small percentage. However, sharing needles and/or syringes with someone who is infected is still a mode of transmission of the HIV virus. In the past, HIV was frequently passed from a mother to her baby during pregnancy and/or birth. Treatments are now available that reduce the chances of a mother passing the virus to her child to one percent.

What is the **purpose** of **vaccination**?

The purpose of vaccination, or immunization, is to artificially induce active immunity so there will be resistance to the pathogen upon natural exposure in the future. Vaccinations are prepared under laboratory conditions from either dead or severely weakened antigens.

Why is a **flu vaccine required** each year?

A new flu vaccine is prepared every year because the strains of flu viruses change from year to year. Nine to 10 months before the flu season begins, scientists prepare a new vaccine made from inactivated (killed) flu viruses. The vaccine preparation is based on the strains of the flu viruses that are in circulation at the time. It includes those A and B viruses expected to circulate the following winter.

Another reason to get vaccinated for the flu every year is that immunity afterwards declines and may be too low to provide protection after one year.

Which **vaccinations** are given during **childhood**?

Childhood vaccinations begin within the first couple of months after birth. The recommended childhood immunization schedule for children in the United States includes: hepatitis B, diphtheria, tetanus, pertussis (DTaP), Haemophilus influenzae type b (Hib), poliovirus, measles, mumps, rubella (MMR), varicella (chickenpox), pneumococcal, and influenza.

How does a **live, attenuated vaccine** differ from an **inactivated (killed) vaccine**?

Live, attenuated, vaccines contain a version of living microbes that has been weakened (attenuated) in a laboratory setting so they can no longer cause disease. Since they are very close to the actual infection, they elicit strong cellular and antibody responses. Live, attenuated vaccines often provide lifelong immunity with only one or two doses. Live, attenuated vaccines are usually more successful with viruses than bacteria.

Inactivated (killed) vaccines are generally better for use against bacteria. Scientists produce inactivated vaccines by killing the disease-causing microbe with chemicals, heat, or radiation. Most inactivated vaccines, however, stimulate a weaker immune system response than do live vaccines. Therefore, it may take several additional doses, called booster shots, to maintain a person's immunity to a particular bacterium.

What are some **diseases** which are **preventable by vaccination**?

Many diseases, such as polio and smallpox, have been eradicated due to vaccination programs. Other diseases, both bacterial and viral, which are preventable by vaccination include:

- Anthrax
- Bacterial meningitis
- Chickenpox
- Cholera
- Diphtheria
- Haemophilus influenzae type b (Hib)
- Hepatitis A
- Hepatitis B
- Influenza
- Measles
- Mumps
- Pertussis (whooping cough)
- Pneumococcal pneumonia
- Polio
- Rabies
- Rubella (German measles)
- Tetanus
- Yellow fever

Is there a **vaccine** to prevent **HIV**?

Researchers are working to develop vaccines to prevent HIV in HIV-negative individuals. They are also trying to develop therapeutic vaccines for HIV-positive individuals to improve their immune system. There is no projected date when a vaccine will be approved for use.

How do physicians determine whether a **donor** and **recipient** are **compatible**?

Researchers have developed a tissue typing technique to determine the compatibility between organ donors and organ recipients. Each individual has a unique set of cell and tissue "markers" called HLA antigens. George Snell (1903–1996), Jean Dausset (1916–) and Baruj Benacerraf (1920–) shared the Nobel Prize in Physiology or Medicine in 1980 for their research on identifying and understanding the genetic structure of cell and tissue markers. The more similar the cell and tissue markers are between donor and recipient, the less likely there will be tissue rejection or graft-versus-host disease.

ALLERGIES

What is an **allergic reaction**?

An allergic reaction is a reaction to a substance that is normally harmless to most other people. Allergens, the antigens that induce an allergic reaction, may be foods, medications, plants or animals, chemicals, dust, or molds.

Which **antibody** is responsible for most **allergic reactions**?

Immunoglobulin E (IgE) is responsible for most allergic reactions. Each type of IgE is specific to a particular allergen. When exposed to an allergen, IgE antibodies attach themselves to mast cells (normal body cells that produce histamines and other chemicals) or basophils. When exposed to the same allergen at a later time, the individual may experience an allergic response when the allergen binds to the antibodies attached to mast cells, causing the cells to release histamine and other inflammatory chemicals.

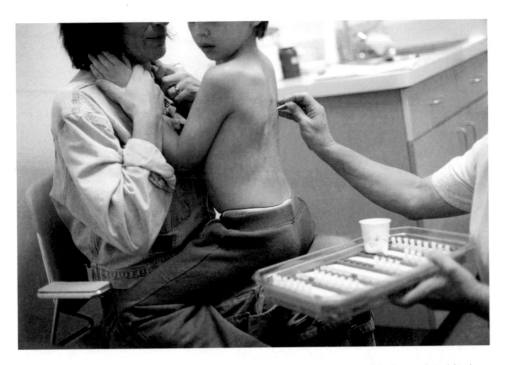

A child is tested for allergic reactions in a doctor's office. The skin is exposed to numerous possible allergens that might trigger an allergic response. © iStockphoto.com/Slobo.

Are **allergies** an **inherited** medical **disorder**?

Hypersensitivity to specific allergens is not inherited. However, the predisposition to developing allergies may be inherited in many people. Studies have found that if neither parent suffers from allergies, the chances of a child developing allergies is only 10 to 20 percent. If one parent has allergies, the chances increase to 30 to 50 percent and if both parents have allergies, the chances increase to 40 to 75 percent. One explanation for this is in the ability to produce higher levels of IgE in response to allergens. Individuals who produce more IgE will develop a stronger allergic sensitivity.

How do **immediate allergic reactions** differ from **delayed allergic reactions**?

Immediate allergic or hypersensitive reactions are mediated by mast cells. The reactions occur within minutes following contact with an allergen. Inhaled or ingested allergens usually cause immediate allergic reactions.

Delayed allergic or hypersensitive reactions are mediated by T cells. The reaction occurs hours and possibly even days after contact with the allergen. Contact on the skin with an allergen is more likely to cause a delayed allergic reaction.

What are **common allergic reactions**?

Allergic reactions include a variety of symptoms. Some common allergic reactions are allergic rhinitis, or "hay fever"; allergic conjunctivitis (an eye reaction); asthma;

atopic dermatitis (skin reactions); urticaria, also known as hives; and severe systemic allergic reactions such as anaphylaxis.

What is **anaphylactic shock**?

Anaphylactic shock is a severe, potentially fatal systemic allergic reaction. It usually involves several organ systems, including the skin, respiratory tract, and gastrointestinal tract. Treatment includes an injection of adrenaline.

What are the **most common allergens**?

The most common allergens are pollen and dust mites.

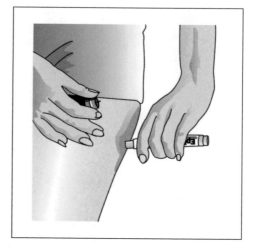

An epinephrine pen (or epipen, for short) can be used to quickly fend off severe allergic reactions, but a doctor should still be consulted as soon as possible. (From Smeltzer, S. C., and Bare, B. G. *Textbook of Medical-Surgical Nursing*. 9th Ed. Philadelphia: Lippincott, Williams & Wilkins, 2000.)

What are some common **food allergies**?

Common food allergy triggers are the proteins in cow's milk, eggs, peanuts, wheat, soy, fish, shellfish, and tree nuts.

How many children suffer from **food allergies**?

Children are more likely to suffer from food allergies than adults. Experts estimate food allergy occurs in six to eight percent of children four years of age or under. Many children lose their sensitivity to problem foods over time.

Why are some **individuals allergic** to **cats** but not to **dogs**?

Individuals allergic only to cats have the IgE antibody specific to cat dander. Allergic reactions to dogs involve a separate, distinct antibody.

Which **drug** causes the **most allergic reactions**?

Penicillin is a common cause of drug allergy. One research study found that approximately seven percent of normal volunteers react to penicillin allergy skin tests (IgE antibodies). Anaphylactic reactions to penicillin occur in 32 of every 100,000 exposed patients.

Why do some individuals experience **allergic reactions** after visits to **doctors and dentists**?

Some individuals are allergic to latex, a component of most rubber gloves. They may experience skin rashes, hives, eye tearing and irritation, wheezing, and itching of the skin when exposed to latex. The best therapy for latex allergy is to avoid products containing latex. Therefore, it is important to notify health care providers if you suffer from latex allergy so they may use non-latex products.

What are "**epipens**"?

Injections of epinephrine are often used to treat systemic, life-threatening allergic reactions. Epipens are needle and syringe medical devices that deliver a premeasured, single-dose of epinephrine as a self-delivered automatic injection.

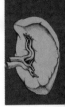

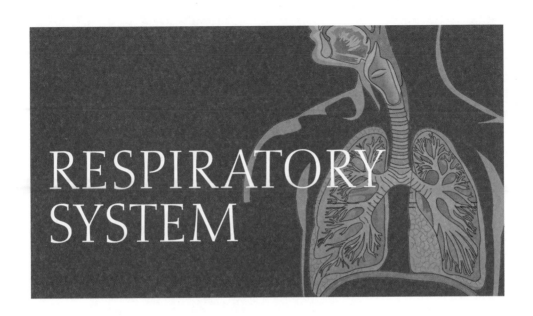

RESPIRATORY SYSTEM

INTRODUCTION

What are the **major functions** of the **respiratory system**?

The major functions of the respiratory system are:

1. Gas exchange: the respiratory system allows oxygen in the air to enter the blood and carbon dioxide to leave the blood and enter the air. The cardiovascular system transports oxygen from the lungs to the cells of the body and carbon dioxide from the cells of the body to the lungs.

2. Regulation of blood pH, which can be altered by changes in blood carbon dioxide levels.

3. Voice production as air moves past the vocal cords to make sound and speech.

4. Olfaction, or the sense of smell, occurs when airborne molecules are drawn into the nasal cavity.

5. Innate immunity, providing protection against some microorganisms by preventing their entry into the body or by removing them from respiratory surfaces.

What are the **two divisions** of the **respiratory system**?

The respiratory system is divided into the upper respiratory system and the lower respiratory system. The upper respiratory system includes the nose, nasal cavity, and sinuses. The lower respiratory system includes the larynx, trachea, bronchi, bronchioles, and alveoli.

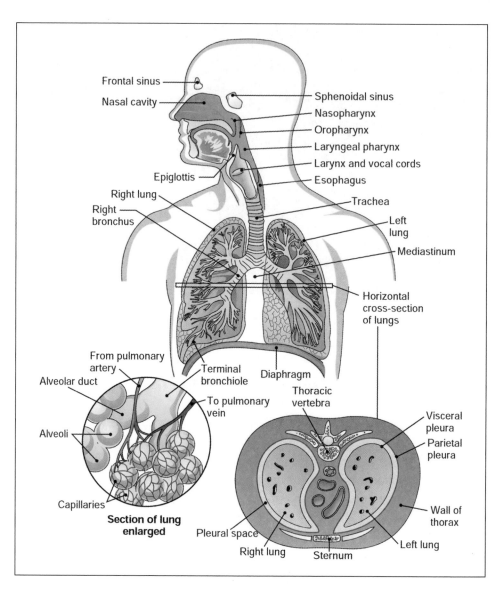

The respiratory system. (From Cohen, B. J., and Wood, D.L. *Memmler's The Human Body in Health and Disease*. 9th Ed. Philadelphia: Lippincott, Williams & Wilkins, 2000.)

What is the basic difference between **respiration** and **cellular respiration**?

Respiration is the entire process of gas exchange between the atmosphere and the body cells. Cellular respiration is the process of oxygen utilization and carbon dioxide at the cellular level.

Why do we **die** within minutes if we are **deprived of oxygen**?

We die within minutes of being deprived of oxygen because each of the trillions of cells in the human body requires oxygen to survive.

What are **non-respiratory air movements**?

Air movements that occur in addition to breathing are called non-respiratory movements. They are used to clear air passages, as in coughing or sneezing, or to express emotions, as in laughing or crying. Also included in this category are hiccupping and yawning, as well as speech, where air is forced through the larynx, causing the vocal cords to vibrate. In speech, words are formed by lips, the tongue, and the soft palate.

STRUCTURE AND FUNCTION

What are the **functions** of the **nose**?

The nose is the primary entry point for air into the respiratory system. It is supported by bone and cartilage. The many hairs that line the nostrils help to filter large particles from air. From the nose, the air is then further filtered and humidified by passages through the sinuses.

How effective are the **structures** in the **nose** at **removing dust, bacteria, and other particles?**

As the cilia of the mucous membrane cells in the nose move, they push a thin layer of mucus and trapped particles, including dust and other small particles such as bacteria, toward the pharynx, where it is swallowed. In the stomach, gastric juice destroys the bacteria and other microorganisms trapped in the mucus. However, some bacterial spores, including those from the bacterium that causes anthrax, are very small and bypass the hairs and mucus in the nose. Spores are able to reach the lungs, where they release a toxin causing inhalation anthrax, and ultimately cause death.

Do **nasal strips** really work to **stop snoring**?

Nasal strips may make a difference for the average snorer, if the snoring originates in the nose. They keep the nostrils wide apart, and by doing so they help prevent snoring and move nitric oxide from the nasal passage into the lungs, improving lung function. Snoring that originates from the base of the tongue or the soft palate is not relieved by nasal strips.

What is a **deviated septum**?

The nasal septum, which is composed of bone and cartilage, divides the nostrils and the nasal cavity into right and left portions. It is usually straight at birth and sometimes bends as the result of a birth injury. The nasal septum is usually straight throughout childhood, but as a person ages the septum tends to bend towards one side or the other. This deviated septum may make breathing difficult by obstructing the nasal cavity.

What causes a **sneeze**?

A sneeze occurs when the cells lining the nose are irritated; the response is a rapid rush of air to force the irritant out of the nasal passages.

Why is the **sneeze reflex** important?

The sneeze reflex functions to dislodge foreign substances from the nasal cavity. Sensory receptors detect the foreign substances and stimulate the trigeminal nerves to the medulla oblongata, where the reflex is triggered. During the sneeze reflex, some air from the lungs passes through the nasal passages, although a significant amount passes through the oral cavity.

What causes a **nosebleed**?

The inside of the nose is heavily supplied with blood vessels. A blow to the nose, excessive nose blowing, infection, allergies, clotting disorders, or hypertension can cause a nosebleed.

Where are the **sinuses** located?

The sinuses are cavities located in the bones surrounding the nasal cavity. The sinuses lighten the skull, and as air flows through the cavities it is warmed and humidified. The nasal cavities are also important in voice production.

What are the functions of the **paranasal sinuses**?

The paranasal sinuses are membrane-linked, air-filled cavities in the bones of the skull that are connected to the nose by passageways. The four paranasal sinuses and the bones in which they are located are the frontal, ethmoid, maxillary, and sphenoid. These cavities lighten the weight of the bones and add resonance to the voice.

Why is **crying** often accompanied by a **runny nose**?

A nasolacrimal (from the Latin, meaning "nose" and "tear") duct leads from each eye to the nasal cavity. Excessive secretion of tears in the eyes is drained into the nose, often producing a watery flow from the nose at times of crying.

Where is **cold air warmed** as it is **breathed** into the body?

Cold air is warmed in the nasal cavity, where the blood vessels in this large surface area provide the capacity for warming inhaled air. As heat leaves the blood in this extensive network of blood vessels, the air is warmed to body temperature.

Why should you not inhale **super-cooled air**?

Inhaled super-cooled air decreases the fluidity of the mucus that lines the respiratory passageways. Super-cooled air can even cause ice crystals to form in the nasal mucosa.

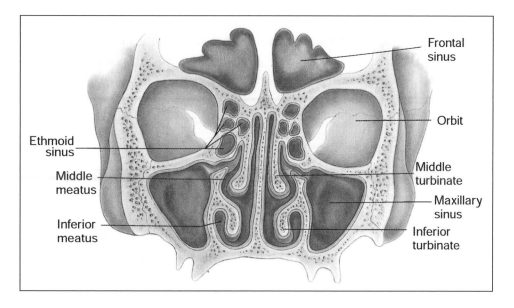

The sinuses are located behind the nose and around the nasal cavity. (From Bickley, L. S., and Szilagyi, P. *Bates' Guide to Physical Examination and History Taking*. 8th Ed. Philadelphia: Lippincott, Williams & Wilkins, 2003.)

What causes a **sinus headache**?

A painful sinus headache can result from blocked drainage of the nasal sinuses caused by an infection or allergic reaction.

What is the **pharynx**?

The pharynx, commonly called the throat, is connected to the nasal cavity and to the oral cavity. It serves as a passageway for food traveling from the oral cavity to the esophagus and for air traveling from the nasal cavity to the larynx. The pharynx has three divisions: 1) the nasopharynx, 2) the oropharynx, and 3) the laryngopharynx.

What causes a **sore throat**?

Most sore throats are caused by viruses, the same germs that cause colds and influenza (the flu). Some sore throats are caused by bacteria such as *Streptococcus*. Other common causes of a sore throat include allergies, dryness of indoor air, pollution and other irritants, muscle strain in the throat, acid reflux disease, HIV infection, and oral tumors.

What is the **larynx**?

The larynx is the passage located between the pharynx and trachea that houses the vocal cords. After air leaves the pharynx, it enters the larynx. The larynx is a cartilaginous structure that encloses the glottis. It is formed by three cartilages: the thyroid, cricoid, and the epiglottic. In addition to forming the passageway for air to travel from the pharynx to the rest of the respiratory tract, the larynx is the site of most sound production.

237

What is the Heimlich maneuver?

The Heimlich maneuver is used in situations where a person has a foreign object lodged in the larynx or trachea. It is often called a "café coronary" because it occurs when a person is choking on a piece of food. When using the Heimlich maneuver, compression is applied to the victim's abdomen just below the diaphragm. This elevates the diaphragm and usually generates enough pressure to push out the foreign object.

What is the **Adam's apple**?

The Adam's apple is a bulge in the throat that is due to the prominent surface of the thyroid cartilage. It is more visible in men because the male larynx is larger than the female larynx and because its prominent part in women is covered by fatty tissue.

Why is it dangerous to **talk while eating**?

If a person talks while eating, food may be inhaled into the lungs. Normally, after food is swallowed, it passes into the pharynx and then into the esophagus. Food is prevented from entering the larynx (the passageway to the lungs) by the epiglottis, a spade-shaped cartilage flap that covers the pharynx. If food does enter the larynx, a cough reflex is usually initiated, although food may lodge in the larynx, causing a blockage of the airway.

What is a **bronchoscopy**?

A bronchoscopy is a direct visual examination of the larynx and airways through a long, flexible viewing tube. A bronchoscope can be inserted through the mouth or nose and extended into the lungs. It can also be used to collect tissue and fluid samples. Bronchoscopy can help a physician make a diagnosis and treat certain medical conditions.

What is **laryngitis**?

Laryngitis is an inflammation of the larynx. It can be caused by smoking, exposure to irritants, or an infection. Because the larynx is the site of sound production, laryngitis usually results in hoarseness or inability to produce audible sounds.

What makes the **trachea flexible**?

The trachea, or windpipe, is a flexible, cylindrical tube approximately 1 inch (2.5 centimeters) in diameter and 5 inches (12.5 centimeters) in length. Within the tracheal wall there are about 20 C-shaped pieces of cartilage, stacked one above the other. The remaining part of the tracheal ring is composed of smooth muscle and connective tissue. This soft tissue allows the nearby esophagus to expand as food moves through it into the stomach.

What is the difference between a **tracheotomy** and a **tracheostomy**?

A tracheotomy is the surgical opening of the trachea. This may be necessary if the trachea becomes occluded through inflammation, excessive secretion, trauma, or aspiration of a foreign object. This procedure may be performed to create an emergency opening into the trachea so that ventilation can still occur. A tracheostomy involves the insertion of a tube into the trachea to permit breathing and to keep the passageway open.

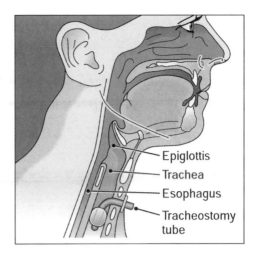

Epiglottis

Trachea

Esophagus

Tracheostomy tube

A tracheostomy involves opening an airway through the throat in an emergency when a patient cannot breath normally. (From Cohen, B. J. *Medical Terminology*. 4th Ed. Philadelphia: Lippincott, Williams & Wilkins, 2003.)

What is the **respiratory tree**?

The term respiratory tree refers to the upside-down tree configuration of the respiratory pathways. It follows this pattern from large branch to smaller branch: trachea → right and left pulmonary bronchi → secondary bronchi → tertiary bronchi → bronchioles → pulmonary lobules → alveoli.

Are both of the **lungs identical**?

The lungs are cone-shaped organs in the thoracic cavity. The right lung consists of three lobes (right superior lobe, right middle lobe, and right inferior lobe) while the left lung has only two lobes (left superior lobe and left inferior lobe) and is slightly smaller than the right lung. Although relatively large, each lung weighs only 1 pound (2.2 kilograms).

What is the **surface area** of the **lungs**?

An adult lung has approximately 300 million alveoli. The total surface area of the lungs is approximately the size of a tennis court.

What is the **consistency** of the **lungs**?

The consistency of the lungs, when inflated, is very porous, much like whipped gelatin.

How **thick** are the **alveoli**, or air sacs, in the lungs?

Each alveoli measures approximately 0.000004 inches (0.00001 centimeters) thick.

What is the essential **relationship** between the **heart** and **lungs**?

The teamwork of the heart and lungs ensures that the body has a constant supply of oxygen for metabolic activities and that the major waste product of metabolism, carbon dioxide, is continuously removed. This occurs through the pulmonary circulation, with the heart supplying blood that has moved through the body to the lungs.

How do the **lungs connect** to the **heart**?

The lungs connect to the heart through blood vessels. The pulmonary artery delivers deoxygenated blood to the lungs from the right ventricle and the pulmonary vein delivers oxygenated blood to the left atrium of the heart.

What **prevents** the lobes of the **lungs** from **rubbing** against each other?

The lobes of the lungs do not normally rub against each other due to the presence of pleural fluid. Pleural fluid is secreted by the membranes that cover the lungs. It reduces the friction between the lobes of the lung and the body cavity.

How **rapidly** is **blood pumped** through the **lungs**?

Once every minute, the heart pumps the body's entire blood supply through the lungs.

What is **pleurisy**?

Pleura is a thin, transparent, two-layered membrane that covers the lungs and also lines the inside of the chest wall. The layer that covers the lungs is in close contact with the layer of the chest wall. Between these two thin layers is a small amount of fluid that lubricates each one as they slide over one another with each breath. Pleurisy is a painful condition which results when there is an abnormal amount of fluid between these two layers. The inflammation from this condition results in friction during inhalation and exhalation.

What is **pneumothorax**?

Pneumothorax refers to the presence of air in the pleural cavity. It occurs when air leaks from the lungs and gets between the lung and the chest wall. It is also known as a collapsed lung and can result from a penetrating chest injury or changes in pressure during diving, flying, or stretching.

Can **lungs** be **transplanted**?

Lung transplantation is surgery used to replace one or both diseased lungs with a healthy lung or lungs. It is only recommended if the patient has end-stage pulmonary disease that cannot be treated any other way. Examples of end-stage pulmonary diseases include emphysema, cystic fibrosis, sarcoidosis, and pulmonary fibrosis. Survival rates for lung transplants are as high as 80 percent at one year and 60 percent at four years after surgery.

What is **COPD**?

COPD (chronic obstructive pulmonary disease) refers to a group of diseases, primarily bronchitis and emphysema. Bronchitis is a chronic inflammation of the bronchi caused by irritants such as cigarette smoke, air pollution, or infections. The inflammation results in swelling of the mucous membrane lining the bronchi, increased mucus production, and decreased movement of mucus by cilia. Emphysema usually follows bronchitis and involves the destruction of the alveolar walls. Ultimately, the lungs lose their elasticity and become less efficient.

What is **emphysema**?

Emphysema is a progressive disease that destroys alveoli, particularly the walls. As a result, groups of small air sacs combine to form larger chambers that have significantly less surface area. This reduces the volume of gases that can be exchanged through the cell membranes.

Smoking has terrible consequences on the lungs, throat, heart, and overall health in general. © iStockphoto.com/Vikram.

What is a **cough**?

A cough is a sudden, explosive movement of air to clear material from the airways.

How does **smoking affect** the **lungs**?

Tobacco smoking, especially cigarettes, is linked to lung cancer, which is the most common cause of cancer death in most countries. Eighty percent of lung cancer cases are due to cigarette smoking. The toxins from cigarettes damage the alveoli so they leak and pop, causing emphysema. Lung cancer usually develops in the epithelium of the airways, spreads through the walls of the airways, and enters the bloodstream and lymphatic system. These changes are usually not detected until secondary growths or other symptoms occur. Only 13 percent of lung cancer patients live as long as five years after the initial diagnosis.

How is **smoking** related to **life expectancy**?

For each pack of cigarettes smoked per day, life expectancy is decreased by seven years. A two-pack-a-day smoker will likely die at 60, compared to 74 for a nonsmoker in the United States.

What is **pulmonary fibrosis**?

Pulmonary fibrosis occurs when chronic inflammation of the lungs results in the replacement of normal, elastic lung tissue with nonelastic scar tissue. Exposure to coal dust, asbestos, and silica are the most common causes of pulmonary fibrosis.

What is an iron lung?

An iron lung is a large machine that is about 7 feet (2.13 meters) long and weighs several hundred pounds. It is used to allow a person to breathe if normal muscle control has been lost or temporarily diminished. The iron lung was invented in 1929 by Dr. Phillip Drinker (1894–1972) and Louis Agassiz Shaw (1886–1940) to help victims of paralytic poliomyelitis. The iron lung is still in use on a limited basis. In the United States, 30 to 40 people still depend on the device.

What is **pneumonia**?

Pneumonia refers to an infection (viral or bacterial) within the lungs. The resulting inflammation from the infection causes fluid accumulation and difficulty breathing. According to the Centers for Disease Control and Prevention, there were 61,472 deaths attributed to pneumonia/influenza making them the eighth leading cause of death in the United States in 2004.

RESPIRATION AND BREATHING

What are the **two phases** of **breathing**?

Breathing, or ventilation, is the process of moving air into and out of the lungs. The two phases are: 1) inspiration, or inhalation; and 2) expiration, or exhalation. Inspiration is the movement of air into the lungs, while expiration is the movement of air out of the lungs.

What is the **respiratory cycle**?

The respiratory cycle consists of one inspiration followed by one expiration. The volume of air that enters or leaves during a single respiratory cycle is called the tidal volume. Tidal volume is typically 500 milliliters, meaning that 500 milliliters of air enters during inspiration and the same amount leaves during expiration.

How does **forced breathing** differ from **quiet breathing**?

Forced breathing involves active inspiratory and expiratory movements. During forced breathing, the accessory muscles assist with inhalation. Exhalation involves contraction of the internal intercostal muscles. The abdominal muscles are involved during the maximum levels of forced breathing. Contraction of the abdominal muscles compresses the abdominal contents, pushing them up against the diaphragm and further reducing the volume of the thoracic cavity. Inhalation during quiet breathing involves contraction of the diaphragm and external intercostals muscles, but exhalation is a passive process.

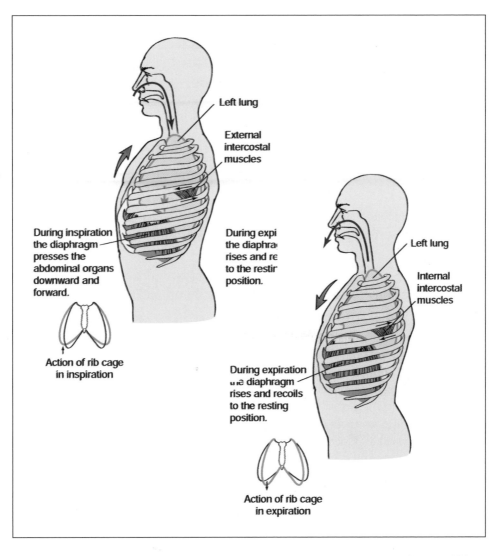

Left lung

External
intercostal
muscles

During inspiration
the diaphragm
presses the
abdominal organs
downward and
forward.

During expi
the diaphra
rises and re
to the restir
position.

Left lung

Internal
intercostal
muscles

Action of rib cage
in inspiration

During expiration
the diaphragm
rises and recoils
to the resting
position.

Action of rib cage
in expiration

Respiration and inspiration. (From Weber, J., R.N., Ed.D., and Kelley, J., R.N., Ph.D., *Health Assessment in Nursing*. 2nd Ed. Philadelphia: Lippincott, Williams & Wilkins, 2003.)

What is the **diaphragm**?

The diaphragm is a muscular separation between the thoracic and abdominal cavities and is the principal muscle involved in inspiration. The contractions of the diaphragm enlarge the thoracic cavity, allowing inspiration or inhaling to move air into the lungs. During expiration, the diaphragm returns to its original position.

Where are **gases** actually **exchanged** in the lungs?

The alveoli are the structures that carry out the vital process of exchanging gases (oxygen and carbon dioxide) between the air and the blood. The tiniest passageways in the lungs are the alveoli. Each pulmonary lobule ends in an alveoli, which are then interconnected with one another. There are approximately 150,000,000 alveoli in each lung.

How does the body introduce oxygen to the blood and where does this happen?

Blood entering the right side of the heart (right atrium) contains carbon dioxide, a waste product of the body. The blood travels to the right ventricle, which pushes it through the pulmonary artery to the lungs. In the lungs, the carbon dioxide is removed and oxygen is added to the blood. Then the blood travels through the pulmonary vein carrying the fresh oxygen to the left side of the heart. It first enters left atrium, where it goes through a one-way valve into the left ventricle, which must push the oxygenated blood to all portions of the body (except the lungs) through a network of arteries and capillaries. The left ventricle must contract with six times the force of the right ventricle, so its muscle wall is therefore twice as thick as that of the right ventricle.

How does external respiration differ from internal respiration?

External respiration is the exchange of gases between the alveoli and lung capillaries. Oxygen diffuses from the alveoli into the blood, while carbon dioxide moves from the blood in the alveoli. Internal respiration, in contrast, is the exchange of gases in body tissues.

Why is carbon monoxide deadly to breathe?

Carbon monoxide is a colorless, odorless gas that has the unique ability to compete with oxygen for binding sites on the hemoglobin molecule. Carbon monoxide binds to the iron in hemoglobin in red blood cells about 200 times as readily as does oxygen, and it tends to stay bound. As a result, hemoglobin bound to carbon monoxide can no longer transport oxygen. Prolonged exposure to carbon monoxide results in carbon monoxide poisoning, with symptoms including nausea, headache, and eventually unconsciousness. If left untreated, death may result. Carbon monoxide is commonly emitted by automobiles and fuel-fired space heaters.

What is **hyperbaric oxygen therapy**?

Hyperbaric oxygen therapy is when a patient is given 100 percent oxygen gas at two to three atmospheres pressure to breathe for varying lengths of time. It is used to treat carbon monoxide poisoning, decompression sickness, severe traumatic injury, infections that could lead to gas gangrene, and other conditions. Normal oxygen concentration is 0.3 milliliters oxygen per 100 milliliters of blood, but breathing 100 percent oxygen at a pressure of three atmospheres raises the plasma concentration to about 6 milliliters of oxygen per 100 milliliters of blood.

What are the "**bends**"?

This term actually refers to a condition known as decompression sickness. It occurs when a diver ascends to the surface too quickly after swimming at significant depths. When this happens, the dissolved nitrogen gas in the tissues literally bubbles out of the tissues and into body fluids. The resulting gas bubbles can cause pain in joints, bones, and muscles.

What are **respiratory acidosis** and **respiratory alkalosis**?

Respiratory acidosis results when the respiratory system is unable to eliminate adequate amounts of carbon dioxide, which then accumulates in the circulatory system, causing the pH of the body fluids to drop. Respiratory alkalosis results from hyperventilation, such as might occur in response to stress.

What is **hyperventilation**?

Hyperventilation is abnormally deep and prolonged breathing.

What **muscles** help in **breathing**?

Breathing is caused by the actions of the muscles between the ribs, the external intercostal muscles, and the diaphragm. When air is breathed in, the intercostal muscles move the ribs upward and outward, and the diaphragm is pushed downward, thus taking air into the expanded lungs. If a person needs to take a deeper than normal breath, the diaphragm and external intercostal muscles contract more forcefully. Additional muscles, such as the pectoralis minor and sternocleidomastoid, can also pull the thoracic cage further upward and outward, enlarging the thoracic cavity and decreasing internal pressure. Breathing out is a passive process. The intercostal muscles return to their resting position, returning the thoracic cage to its original size and expelling air from the lungs as a result.

What is **pulmonary surfactant**?

Pulmonary surfactant is an oily substance (a mixture of lipoprotein molecules) produced by the cells lining the alveoli. Since all parts of the respiratory system are coated with a thin watery layer, the surfactant coats this watery layer and reduces the surface tension within the alveoli. This helps keep the alveoli open for gas exchange.

What is **respiratory distress syndrome (RDS)**?

In a developing embryo pulmonary surfactant is not produced in sufficient quantities until the seventh month of gestation. As the fetus matures, the amount of pulmonary surfactant increases. In premature infants, respiratory distress syndrome (RDS), or hyaline membrane disease, is caused by too little surfactant. This is common, particularly for infants delivered before the seventh month of pregnancy.

How is **air cleaned** before reaching the **lungs**?

There are many ways by which air is cleansed before it reaches the lungs. As air is inhaled through the nose, the outer part of the nostril has visible hairs (vibrissae) that filter larger particles. In addition, the nasal cavity has a mucous lining that traps smaller particles or microorganisms. The mucus then flows toward the pharynx, where it is swallowed; any trapped microbes are usually killed when exposed to the acidic conditions in the stomach. The cells lining the respiratory system have cilia (tiny, oar-like appendages) on the surface that act as an additional filter to trap debris.

What part of the **brain** controls **respiratory rate**?

The major center of respiratory activity is the medullary respiratory center in the medulla oblongata. It consists of two dorsal respiratory groups and two ventral respiratory groups. The dorsal groups stimulate contraction of the diaphragm and the ventral groups stimulate various groups of muscles. In addition, the pontine respiratory group, located in the pons, functions in switching between inspiration and expiration.

How is **breathing** related to **age**?

As we get older, we tend to breathe slower, as shown by the table below. Gender also is a factor in breathing rates.

Age in years	Breaths per minute
Infant	40–60
5	24–26
15	20–22
25 (male)	14–18
25 (female)	16–20

After the age of 25, breathing rates level off. In the average lifetime, an individual takes more than 600 million breaths of air.

What is a **normal respiratory rate**?

The average person breathes about 16 times each minute and takes in one pint of air (0.5 liters) with each breath. In an average lifetime, we breathe over 75 million gallons (2.85 trillion liters) of air.

How is the amount of **air moved** each minute **calculated**?

The respiratory minute volume is the amount of air moved each minute. It is calculated by multiplying the respiratory rate (12 breaths per minute on average) by the tidal volume (500 milliters per minute) or 6,000 milliliters (6 liters) per minute.

How much **air** is **required** by the body for **general activities**?

The body requires 8 quarts (7.6 liters) of air per minute when lying down, 16 quarts (15.2 liters) when sitting, 24 quarts (22.8 liters) when walking, and 50 quarts (47.5 liters) when running.

What is the **volume** of air in the **lungs**?

The total volume of air the lungs of an average young adult can hold, also called total lung capacity (TLC), is 5,800 milliliters. This is the combination of the vital capacity (VC) (4,600 milliliters) and the residual volume (RV) (1,200 milliliters). The vital capacity is the maximum volume of air that can be exhaled after taking the deepest breath possible. The residual volume is the volume of air that remains in the lungs even after maximal expiration.

Why is it more difficult to **breathe** at **high altitudes**?

It is difficult to breathe at high altitudes because there is less oxygen available in the atmosphere. If the concentration of oxygen in the alveoli drops, the amount of oxygen in the blood drops. At altitudes of 9,843 feet (3,000 meters) or more, people often feel lightheaded, especially if they are exercising and placing extra demands on the cardiovascular and respiratory systems.

How does **lung capacity** differ in **men** and **women**?

While tidal volume remains constant between men and women at 500 milliliters, there are differences in other lung volumes and capacities. Total lung capacity is 4,200 milliliters for women and 6,000 milliliters for men. The residual volume of air is approximately 1,100 milliliters for women and 1,200 milliliters for men. The inspiratory reserve volume is approximately 1,900 milliliters for women and 3,300 milliliters for men. The expiratory reserve volume is about 700 milliliters in women and 1,000 milliliters in men.

The pitch of men's voices is typically lower than women's because their vocal chords are longer. © iStockphoto.com/Robert Kohlhuber.

SOUND PRODUCTION

How do human beings **create sound**?

Air passing over the vocal cords causes them to vibrate from side to side generating sound waves. The frequency of vibration may fluctuate between 50 hertz in a deep bass to 1,700 hertz in a high soprano.

Why is a **man's voice** usually **lower** than a **woman's voice**?

The pitch of the voice—how high or low it sounds—depends on the length, tension, and thickness of the vocal cords. Because males have longer vocal cords of up to 1 inch (2.54 centimeters) in length, the male voice is deeper in pitch, while women and children with shorter cords have higher-pitched voices. Vocal cords in women average 0.167 inches (0.42 centimeters) in length. Testosterone is the hormone that is responsible for the increase of length of male vocal cords during puberty.

Why does the **voice** sound **different** when someone has a **cold**?

When the nasal cavity and paranasal sinuses are filled with mucus instead of air, the quality of the sound produced changes.

What is **phonation**?

Sound production in the larynx is called phonation. It is one component of speech production that also requires articulation, or the modification of those sounds by other anatomical structures. These structures include the pharynx, oral cavity, nasal cavity, and paranasal sinuses. This combination determines the particular and distinctive sound of an individual's voice.

What determines the **pitch** of the sound produced by the **voice**?

The pitch of the sound produced depends on the diameter, length, and tension in the vocal cords. The diameter and length are directly related to the size of the larynx. The tension is controlled by the contraction of voluntary muscles associated with various cartilages of the larynx.

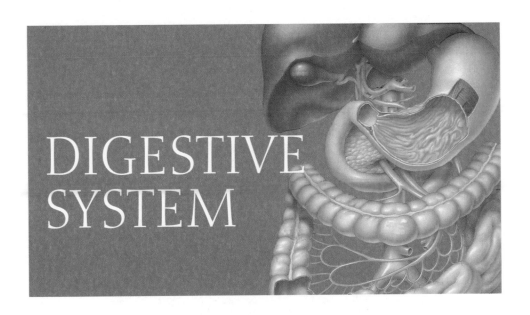

DIGESTIVE SYSTEM

INTRODUCTION

What is the **function** of the **digestive system**?

The purpose of digestion is to process food into molecules that can be absorbed and utilized by the cells of the body as a source of energy for growth and reproduction.

What are the **steps** in the **digestive process**?

There are five major steps in the process of digestion.

1. Ingestion: the eating of any food
2. Peristalsis: the involuntary muscle contractions that move the ingested food through the digestive tract
3. Digestion: the conversion of the food molecules into nutrients that can then be used by the body
4. Absorption: the passage of the nutrients into the bloodstream and/or lymphatic system to be used by the body's cells
5. Defecation: the elimination of the undigested and unabsorbed ingested materials

What are the **major organs** of the **digestive system**?

The digestive system consists of the upper gastrointestinal tract, the lower gastrointestinal tract, and the accessory organs. The organs of the upper gastrointestinal tract are the oral cavity, esophagus, and stomach. The organs of the lower gastrointestinal tract are the small intestine and large intestine (also called the colon). The accessory organs are the salivary glands, the liver, gall bladder, and pancreas.

251

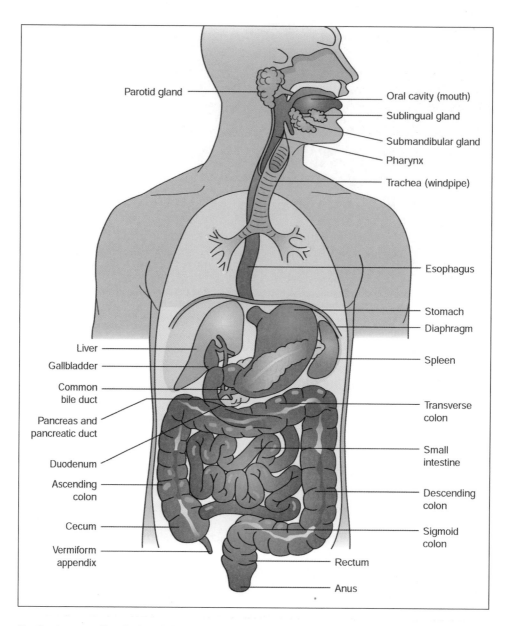

The digestive system. (From Smeltzer, S. C., and Bare, B. G. *Textbook of Medical-Surgical Nursing.* 9th Ed. Philadelphia: Lippincott, Williams & Wilkins, 2000.)

How **long** is the **digestive tract**?

The digestive tract, also called the alimentary canal, is approximately 30 feet (9 meters) long from the mouth to the anus.

What are some **diagnostic procedures** used to examine the **digestive tract**?

Several diagnostic tests are available to examine organs of the digestive tract and to determine causes of abdominal pain and disorders that affect the digestive system.

Who performed some of the earliest studies on digestion?

William Beaumont (1785–1853), an army surgeon, performed some of the earliest studies on digestion. In 1822, Alexis St. Martin was accidentally wounded by a shotgun blast. Beaumont began treatment of the wound immediately. St. Martin's recuperation lasted nearly three years, and the enormous wound healed, except for a small opening leading into his stomach. A fold of flesh covered this opening; when this was pushed aside the interior of the stomach was exposed to view. Through the opening, Beaumont was able to extract and analyze gastric juice and stomach contents at various stages of digestion, observe changes in secretions, and note the stomach's muscular movements. The results of his experiments and observations formed the basis of our modern knowledge of digestion.

Some of the commonly performed screening tests are colonoscopy, flexible sigmoidoscopy, endoscopy, upper GI series and lower GI series X-rays, ERCP (endoscopic retrograde cholangiopancreatography), and liver biopsy.

Colonoscopy allows a physician to look inside the entire large intestine. It is used to detect early signs of cancer in the colon and rectum.

Flexible sigmoidoscopy allows a physician to examine the inside of the large intestine from the rectum through the sigmoid or descending colon (the last part of the colon). It is used to detect the early signs of cancer in the descending colon and rectum.

Upper endoscopy allows a physician to look inside the esophagus, stomach, and duodenum (the first part of the small intestine). This procedure is used to discover the reason for swallowing difficulties, nausea, vomiting, reflux, bleeding, indigestion, abdominal pain, or chest pain.

The *upper GI series* uses X-rays to diagnose problems in the esophagus, stomach, and duodenum. Ulcers, scar tissue, abnormal growths, hernias, or areas where something is blocking the normal path of food through the digestive system are visible with the upper GI series.

The *lower GI series* uses X-rays to diagnose problems in the large intestine, including the colon and rectum. Problems such as abnormal growths, ulcers, polyps, diverticuli, and colon cancer may be diagnosed through a lower GI series.

ERCP (endoscopic retrograde cholangiopancreatography) enables a physician to diagnose and treat problems in the liver, gall bladder, bile ducts, and pancreas.

Liver biopsy is performed when other liver function tests reveal the liver is not working properly. It allows a physician to examine a small sample of liver tissue for signs of damage or disease.

253

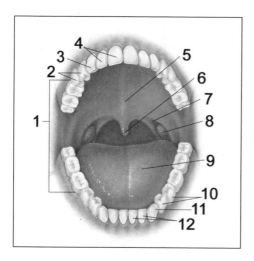

The parts of the mouth include 1 = Molars; 2 = Premolars; 3 = Canines; 4 = Incisors; 5 = Soft palate; 6 = Uvula; 7 = Palatoglossal arch; 8 = Palatine tonsil; 9 = Dorsum of tongue; and complementary teeth on the lower jaw, including 10 = Premolars; 11 = Canines; 12 = Incisors. *Anatomical Chart Co.*

UPPER GASTRO-INTESTINAL TRACT

What is the major **function** of the **upper gastrointestinal tract**?

The upper gastrointestinal tract is the site of food processing. Most mechanical digestion occurs in the upper gastrointestinal tract.

Which structures form the **oral cavity**?

The oral cavity, also called the buccal (from the Latin *bucca,* meaning "cheek") cavity, is formed by the mouth, lips, and cheeks. The teeth, tongue, palate, and salivary glands are associated with the oral cavity.

How many different **types of teeth** are in the mouth?

The three major types of teeth are incisors, cuspids (canines), and molars. All teeth have the same basic structure, consisting of a root, a crown, and a neck. The root is embedded in the socket in the jaw. The crown is the portion that projects up from the gum. The neck, surrounded by gum, forms the connection between the root and the crown.

The different types of teeth perform different functions. The incisors, located at the front of the mouth, are blade-shaped and suited for clipping or cutting. Incisors are important to bite off pieces of food. Located next to the incisors are the cuspids or canines. Their characteristic pointed tips make them suitable for tearing, shearing, and shredding food. Both premolars (also called bicuspids) and molars have flattened crowns with prominent ridges. They are essential for crushing and grinding food.

What is the purpose of **primary teeth**?

Primary teeth, also known as baby, deciduous (from the Latin *deciduus* or *decidere,* meaning to "fall down or off"), temporary, or milk (for their milk-white color) teeth serve many of the same purposes as permanent teeth. They are needed for chewing and they are necessary for speech development. They also prepare the mouth for the permanent teeth by maintaining space for the permanent teeth to emerge in proper alignment. Each individual has 20 primary teeth followed by 32 permanent teeth.

What is the **tongue**?

The tongue is composed mostly of striated muscle. It is divided into two major sections: the oral, or anterior body, part and the pharyngeal, or posterior, part. The oral part is

How much force does a human bite generate?

All the jaw muscles working together can close the teeth with a force as great as 55 pounds (25 kilograms) on the incisors or 200 pounds (90.7 kilograms) on the molars. A force as great as 268 pounds (122 kilograms) for molars has been reported.

covered with small projections called papillae. These papillae give the tongue its characteristic rough texture. There is also a series of taste buds on the tongue. The tongue aids mechanical digestion and is important for sensory input and speech production.

How are the **teeth** and **tongue** involved in **chewing**?

The first stage of mechanical digestion is mastication, or chewing. Initially, the teeth tear and shred large pieces of food into smaller units. The muscles of the tongue, cheeks, and lips help keep the food on the surfaces of the teeth. The tongue then compacts the food into a small round mass of material called the bolus. The salivary glands help lubricate the food with secretions so it is moist.

What is the **composition** of **saliva**?

Nearly 99.5 percent of the total composition of saliva is water. The remaining 0.5 percent consists of ions, such as potassium, chloride, sodium, and phosphates, which serve as buffers and activate the enzymatic activity. An important enzyme in saliva is salivary amylase. It breaks down complex carbohydrates, such as starches, into smaller molecules that can be absorbed by the digestive tract.

How much saliva does a person produce in a day?

Saliva is a mixture of mucus, water, salts, and the enzymes that break down carbohydrates. Awake individuals secrete saliva at a rate of approximately 0.5 milliliters per minute, or an average of 480 milliliters of saliva in a 16-hour waking day. Various activities such as exercise, eating, drinking, and speaking increase salivary volume.

When a **person swallows** solid or liquid food, what prevents it from going down the **windpipe**?

Once food is chewed, voluntary muscles move it to the throat. In the pharynx (throat), automatic, involuntary reflexes take over. The epiglottis closes over the larynx (voice box), which leads to the windpipe. A sphincter at the top of the esophagus relaxes, allowing the food to enter the digestive tract.

What are the actions of **swallowing**?

Swallowing, also called deglutition, involves both voluntary action and involuntary action. During the voluntary phase, the bolus is pushed to the back of the mouth

255

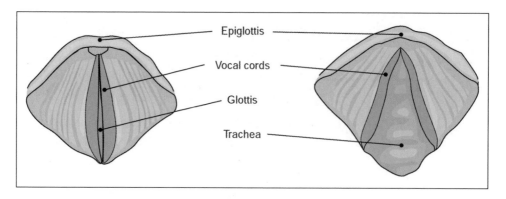

Several structures surround the opening of the throat. (Cohen, B.J., and Wood, D. L. *Memmler's The Human Body in Health and Disease.* 9th Ed. Philadelphia: Lippincott, Williams & Wilkins, 2000.)

by the tongue. The pressure of the bolus on the pharynx activates the swallowing center in the brain. Four steps then occur in rapid succession for the bolus to enter the esophagus.

1. The tongue moves to the roof of the mouth (the hard palate) to prevent food from re-entering the oral cavity.
2. The uvula (the flap of skin at the back of the mouth) moves upward to block the nasal passages.
3. The vocal chords close over the opening of the windpipe or glottis.
4. As the bolus passes into the esophagus, the epiglottis (a flap of cartilaginous tissue) closes over the glottis to ensure food does not enter the respiratory system.

What is **peristalsis**?

Peristalsis is the movement that propels food particles through the digestive tract. Rhythmic waves of smooth muscle contractions perform this action.

What is the role of the **esophagus** in digestion?

The esophagus is a muscular tube approximately 10 inches (25 centimeters) long and 0.75 inches (2 centimeters) in diameter that allows solid food and liquids to pass from the pharynx to the stomach. It is lined with cells that secrete mucus to lubricate the tube and allow for smoother passage of food through the esophagus. Peristaltic action moves the food through the esophagus.

How long does food stay in the esophagus?

Food stays in the esophagus an average of five to nine seconds.

What ensures that **food moves** only **one way** through the esophagus?

A sphincter muscle is found at each end of the esophagus to ensure one-way movement of food. The sphincter at the upper end of the esophagus is the pharyngoesophageal sphincter. The gastroesophageal sphincter, also called the cardiac sphincter, is at the lower end of the esophagus. It is important because it prevents acids from the stomach from being forced into the esophagus and irritating the lining of the esophagus.

Which medical **condition** may occur when the **lower esophageal sphincter** does **not close properly**?

Gastroesophageal reflux disease (GERD) occurs when the lower esophageal sphincter does not close properly and stomach contents leak back, or reflux, into the esophagus. When refluxed stomach acid touches the lining of the esophagus, it causes a burning sensation in the chest or throat called heartburn. The fluid may even be tasted in the back of the mouth, and this is called acid indigestion.

How is **gastroesophageal reflux disease (GERD) treated**?

Lifestyle changes help prevent GERD. Among these are quitting smoking, weight loss, not eating within two to three hours before bedtime, eating more frequent but smaller meals, and avoiding certain foods that may aggravate heartburn, including greasy or spicy foods, coffee, alcohol, chocolate, and tomato products. If lifestyle changes alone do not alleviate the symptoms of GERD, medications, including antacids, acid blockers, or proton pump inhibitors, may be beneficial.

What **functions** are performed by the **stomach**?

The stomach serves several important functions in the digestion of food. It stores ingested food; it is the site of mechanical and enzymatic digestion of ingested food; and it produces and secretes intrinsic factor. Intrinsic factor is a compound secreted by cells of the stomach that facilitates the absorption of vitamin B_{12} in the small intestine.

What are the **four regions** of the **stomach**?

The stomach, which is located directly under the dome of the diaphragm and protected by the rib cage, is a J-shaped organ that has a maximum length of 10 inches (25 centimeters) and a maximum width of 6 inches (15 centimeters). It is divided into four regions: 1) the cardia, 2) the fundus, 3) the body, and 4) the pylorus. Each region is slightly different anatomically. The cardia is located near the gastroesophageal junction. The fundus is the small, rounded part of the stomach located above the gastroesophageal sphincter. The body is the main region of the stomach. It is the area

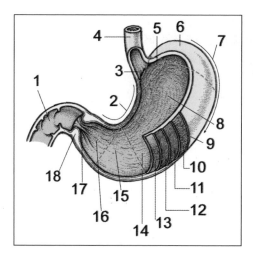

The parts of the stomach: 1 = Duodenum; 2 = Lesser curvature; 3 = Lower esophageal sphincter; 4 = Esophagus; 5 = Cardia; 6 = Fundus; 7 = Greater curvature; 8 = Body; 9 = Serosa; 10 = Muscularis: longitudinal muscle; 11 = Muscularis: circular muscle; 12 = Muscularis: oblique muscle; 13 = Mucosa; 14 = Rugae of mucosa; 15 = Pyloric antrum; 16 = Pylorus; 17 = Pyloric canal; 18 = Pyloric sphincter. (From *Stedman's Medical Dictionary*. 27th Ed. Baltimore: Lippincott, Williams & Wilkins, 2000.)

between the fundus and the "J" shape of the stomach. Most food storage and mixing occur in the body. The pylorus is the bottom curve of the "J" shape. It is located at the junction between the stomach and the small intestine.

How does the **volume** of the **stomach change** from when it is empty to when it is full?

The inner mucous membrane of the stomach contain branching wrinkles called rugae (from the Latin, meaning "folds"). As the stomach fills, the rugae flatten until they almost disappear when the stomach is full. An empty stomach has a volume of only 0.05 quarts (50 milliliters). A full stomach expands to contain 1 to 1.5 quarts (a little less than 1 to 1.5 liters) of food in the process of being digested.

What is **gastric juice**?

Gastric juice is a clear, colorless fluid secreted by specialized cells in the fundus of the stomach. It contains hydrochloric acid, pepsinogen (an inactive proenzyme that is converted to pepsin), mucus, and intrinsic factor. An average of 1.5 quarts (a little less than 1.5 liters) of gastric juice is secreted daily. The major function of gastric juice is to digest protein. The acidity of gastric juice denatures proteins and inactivates most of the enzymes in food. The acidity of gastric juice creates an environment that is unfriendly to many microorganisms ingested with food that may be harmful.

How frequently are the **cells** of the **stomach lining renewed**?

The lining of the stomach sheds about 500,000 cells per minute. It is completely renewed every three days.

What causes a **stomach ulcer**?

Historically, doctors thought that genetics, anxiety, or even spicy foods caused stomach ulcers. While these may worsen the pain of an ulcer, scientists now believe that the gastric ulcer is caused by a bacterium called *Helicobacter pylori*. Researcher Barry J. Marshall (1951–) observed that many ulcer patients had these bacteria present in their systems. In 1984 Marshall designed an experiment to determine whether there was a link between *Helicobacter pylori* and stomach ulcers. He consumed a large amount of the bacteria and waited. After 10 days, he developed ulcers. Marshall shared the 2005 Nobel Prize in Physiology or Medicine with J. Robin Warren (1937–) for their discovery of the bacterium *Helicobacter pylori* and its role in gastritis and peptic ulcer disease.

What is **chyme**?

Chyme is the soupy mixture of partially digested food that forms in the stomach. It is highly acidic.

How does **food leave** the **stomach** and **move** to the **intestine**?

The pyloric sphincter, located between the stomach and duodenum, or small intestine, is never completely closed. Water and other fluids pass continually from the stomach to the duodenum. A small amount of chyme moves through the pyloric sphincter into the duodenum with each peristaltic contraction. The remainder of the chyme is forced back into the pyloric region of the stomach for further mixing.

How long does food remain in the stomach?

Food remains in the stomach for one to three hours as it becomes partially digested and forms chyme.

Does **absorption** of **nutrients** occur in the stomach?

Absorption of nutrients does not occur in the stomach. Most carbohydrates, lipids, and proteins are only partially broken down by the time they leave the stomach and cannot be absorbed. In addition, the lining of the stomach is covered with alkaline cells so the acidic chyme is not in direct contact with the stomach lining. It is relatively impermeable to water. Absorption of nutrients occurs after the chyme has left the stomach.

Are **drugs absorbed** through the **stomach**?

Some drugs, such as aspirin and alcohol, are able to be absorbed through the lining of the stomach. They penetrate the mucous layer of the stomach and enter the circulatory system. Therefore, alcohol in the stomach will be absorbed before nutrients from a meal reach the bloodstream.

Why do **people burp**?

Burping, technically called eructation (from the Latin *ructare,* meaning "belch") is a normal occurrence that results from an abundance of air in the stomach. Nearly a half a quart of air is typically swallowed during a meal. Much of this air is relieved as a burp or belch.

LOWER GASTROINTESTINAL TRACT

What are the major **functions** of the **lower gastrointestinal tract**?

The lower gastrointestinal tract, consisting of the small intestine and large intestine, is the main location where nutrient processing and absorption occurs.

What is the **length** of the human **intestine**?

The small intestine is about 22 feet (7 meters) long. The large intestine is about 5 feet (1.5 meters) long.

How is the **small intestine divided** into segments?

The small intestine is divided into three segments: 1) the duodenum, 2) the jejunum, and 3) the ileum. The duodenum is approximately 10 inches long (25 centimeters) and begins at the junction between the stomach and small intestine at the pyloric sphincter. The jejunum is about 8 feet (2.5 meters) long. The longest segment of the small intestine is the last segment, the ileum. It is 12 feet (3.5 meters) long. Each segment has the same basic anatomical structure and appearance.

Does each **segment** of the **small intestine** have the **same function**?

The final digestive processing of chyme takes place in the duodenum. Once the chyme passes from the stomach to the duodenum, it is mixed with digestive secretions from the liver and pancreas in the final steps in digestion. Most of the absorption of nutrients into the bloodstream and lymphatic system takes place in the jejunum and ileum.

How long does it take for **chyme** to move **through the small intestine**?

It takes an average of one to six hours for chyme to move the length of the small intestine.

How does **chyme move** through the **small intestine**?

Chyme moves through the small intestine by two different types of contractions: peristalsis and segmentation. Peristalsis is the rhythmic contractions that move chyme through the gastrointestinal tract. Segmentation involves localized contractions of small segments of the small intestine. These contractions mix the chyme with the secretions of the small intestine, gall bladder, and pancreas. The nutrients are brought into contact with the microvilli in the small intestine. The chyme is slowly propelled toward the ileocecal valve. Unlike peristaltic contractions, which are directional, the contractions of segmentation are not directional. Therefore, in order to keep the chyme moving downward, the duodenum contracts more frequently than the jejunum or ileum.

What is the purpose of **villi**?

The mucosa layer of the small intestine has many fingerlike projections called villi. Villi (from the Latin, meaning "shaggy hairs"), and the smaller microvilli increase the surface area of the small intestine dramatically, allowing for greater absorption of nutrients. If the small intestine were a smooth tube without villi, it would have a total absorptive area of 3.6 square feet (3,344 square centimeters). The existence of villi effectively increases the absorptive area of the small intestine by a factor of nearly 600 to more than 2,200 square feet (2,043,800 square centimeters).

What is the role of the **small intestine** in **nutrient processing**?

The small intestine is the site of most nutrient processing in the body. The first step is to breakdown the large complex structures of all nutrients, including carbohydrates, lipids, proteins, and nucleic acids, into smaller units. Most absorption of these nutrients also takes place in the small intestine.

What **type of diet** is recommended for individuals with **celiac disease**?

A gluten-free diet is the only treatment for individuals with celiac disease. Celiac disease is an autoimmune digestive disease that damages the small intestine and interferes with absorption of nutrients. The villi in the small intestine are damaged or destroyed whenever sufferers of celiac disease eat products that contain gluten. Gluten is found in wheat, rye, and barley. Once the villi are damaged, they are not able to allow nutrients to be absorbed by the bloodstream, leading to malnutrition.

Which **valve** separates the **small intestine** and the **large intestine**?

The ileocecal valve serves as the boundary between the small intestine and the large intestine. The large intestine frames the small intestine on three sides.

How many distinct **regions** are in the **large intestine**?

The large intestine consists of three distinct regions: 1) the cecum, 2) the colon, and 3) the rectum. The cecum is the first section of the large intestine below the ileocecal valve. The appendix is attached to the cecum. Since the colon (ascending,

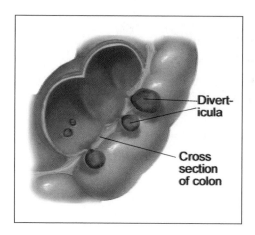

Diverticula

Cross section of colon

Diverticulosis is the formation of painless pouches in the colon. *Anatomical Chart Co.*

transverse, descending, and sigmoid colon) is the largest region of the large intestine, the term "colon" is often applied to the entire large intestine. The rectum (rectum, anal canal, and anus) is the final region of the large intestine and the end of the digestive tract.

What are the **functions** of the **large intestine**?

The large intestine is mostly a storage site for undigested materials until they are eliminated from the body via defecation. Although digestion is complete by the time the chyme enters the large intestine and most absorption has occurred in the small intestine, water and electrolytes are still absorbed through the large intestine.

How does **diverticulosis** differ from **diverticulitis**?

Diverticula are bulging, sac-like pouches in the wall of the colon that protrude outward from the wall of the colon. Diverticula appear most often in individuals over 40 whose diet is low in fiber. In diverticulosis the pouches are present but the individuals do not have any symptoms or discomfort. In diverticulitis, the diverticula are inflamed and often infected when undigested food and bacteria are caught in the diverticula. Patients experience pain, either constipation or increased frequency of defecation, nausea, vomiting, and low-grade fever. In severe cases, surgical removal of the infected area of the colon is necessary. Changing to a high-fiber diet usually relieves the symptoms.

How much **water enters** the **large intestine** daily?

Nearly 95 percent of the water that enters the digestive tract is absorbed by the small intestine. Only 0.5 quarts (0.47 liters) of water enters the large intestine daily. Absorption of water in the large intestine helps to avoid dehydration. Unabsorbed water is excreted with feces.

How long do the **undigested remains** from the digestion process remain in the **large intestine**?

The undigested remains of the digestion process stay in the large intestine for 12 to 36 hours.

Which **gases** are found in **flatus**?

Flatus consists mostly of nitrogen, carbon dioxide, and hydrogen. Small amount of oxygen, methane, and hydrogen sulfide are also a part of flatus (gas).

What is the role of bacteria in the large intestine?

Many bacteria, such as *Escherichia coli* and *Clostridium,* are normally found in the colon. Some bacteria are able to break down undigested fiber into glucose. Bacteria are also able to synthesize certain vitamins, especially vitamins K and B_{12}. The presence of large colonies of beneficial bacteria inhibits the growth of pathogenic bacteria. In addition, the bacteria ferment some of the indigestible carbohydrates, including cellulose, contributing to the production of intestinal gas, or flatus (from the Latin, meaning "blowing").

How much **flatulence** is produced **daily**?

Most individuals produce nearly 1 pint (473 milliliters) of flatus (gas) daily. Passing gas 10 to 20 times per day is considered normal. Diets that consist of an abundance of carbohydrate-rich foods (such as beans) produce greater amounts of gas because a greater amount of undigested carbohydrates enter the large intestine. The mixture of different gases causes the characteristic odor associated with flatulence.

What is **feces**?

Feces (from the Latin *faex,* meaning "dregs") is the remaining portion of undigested food. Approximately 5.3 ounces (150 grams) of feces are produced daily. Feces normally consists of 3.5 ounces (100 grams) of water and 1.7 ounces (50 grams) of solid material. The solid material is composed of fat, nitrogen, bile pigments, undigested food such as cellulose, and other waste products from the blood of intestinal wall.

What gives **feces** its **color**?

The normal brown color of feces is caused by bilirubin. Blood and foods containing large amount of iron will darken the feces. Excessive fat from the diet causes feces to be a more pale color.

What is the **process** of **defecation**?

Defecation is the final step in the digestive process that removes all undigested materials from the body. Defecation involves both voluntary and involuntary actions. When feces fills the rectum, it triggers the defecation reflex. The urge to defecate can be controlled in most individuals except young children and others who have suffered spinal cord injuries.

What is **constipation**?

Constipation may be defined as the passage of small amounts of hard, dry bowel movements. Constipation is the result of too much water being absorbed into the colon. In addition, when the fecal material moves through the colon at a slow rate, more water is absorbed, resulting in hard, dry feces. Diets high in fiber help prevent constipation.

What are some common **causes** of **diarrhea**?

Diarrhea, which is frequent, loose, watery bowel movements, may be caused by infections or other intestinal disorders. Diarrhea may be associated with both bacterial and viral infections. Common bacteria, such as *Campylobacter, Salmonella, Shigella,* and *Escherichia coli,* consumed in contaminated food and/or water will cause diarrhea. Many viruses cause diarrhea, including rotavirus, Norwalk virus, cytomegalovirus, herpes simplex virus, and viral hepatitis. In addition to bacterial and viral infections, parasites, including *Giardia lamblia, Entamoeba histolytica,* and *Cryptosporidium,* may enter the body through food and water and cause diarrhea. Several disorders, including irritable bowel syndrome, inflammatory bowel disease, celiac disease, and side effects of medication may also cause diarrhea. Most cases of diarrhea often resolve themselves without medical intervention. It is important to prevent dehydration by replacing fluids and electrolytes. It is usually recommended to avoid milk products, greasy foods, very sweet foods, and foods that are high in fiber until the diarrhea has subsided. Bland foods may then be slowly reintroduced to the diet.

Which **two diseases** are considered **inflammatory bowel diseases**?

Inflammatory bowel disease (IBD) is the general term for diseases that cause inflammation in the intestines. Crohn's disease and ulcerative colitis belong to the group of illnesses known as IBDs. Crohn's disease may affect any part of the digestive tract, but it most often affects the ileum. Ulcerative colitis occurs only in the inner lining of the colon (the large intestine) and the rectum. Abdominal pain and diarrhea are the most common symptoms of both ulcerative colitis and Crohn's disease. Both diseases are chronic, ongoing diseases, although periods of remission of not uncommon.

How does **irritable bowel syndrome** differ from **inflammatory bowel disease**?

Irritable bowel syndrome is not a disease but a disorder that disrupts the functions of the colon. There is no inflammation of the digestive tract. It is characterized by a group of symptoms, including crampy abdominal pain, bloating, constipation, and diarrhea.

Is the large intestine essential for life?

Since the role of the large intestine is mainly as a storage site for fecal material and the elimination of it from the body, it is not essential for life. Individuals who suffer from colon cancer or other diseases will often have their colon removed. The end of the ileum is brought to the abdominal wall. Food residues are eliminated directly from the ileum into a sac attached to the abdominal wall on the outside of the body. Alternatively, the ileum may be connected directly to the anal canal.

Why is **screening** for **colorectal cancer** important?

Colorectal cancer is the most common cancer of the digestive system. Screening tests are important to diagnose a disease prior to developing symptoms. When detected in the early stage, the five-year survival rate for colorectal cancer is greater than 90 percent. In addition, polyps, which are not malignant, may be removed during a screening procedure, thus avoiding cancer. The screening guidelines suggested by the American Cancer Society for both men and women over the age 50 with average risk for colorectal cancer include:

1. A fecal occult blood test (FOBT) or fecal immunochemical test (FIT) every year, or
2. Flexible sigmoidoscopy every five years, or
3. An FOBT or FIT every year, plus flexible sigmoidoscopy every five years, or
4. Double-contrast barium enema every five years, or
5. Colonoscopy every 10 years

Of the first three options, the combination of FOBT or FIT every year, plus flexible sigmoidoscopy every five years, is preferable.

ACCESSORY ORGANS

Which organs are considered **accessory organs** to the **digestive system**?

The pancreas, liver, and gall bladder are accessory organs in digestion. None of these organs are a part of the digestive tract that begins at the mouth and ends at the anus, but they contribute important chemicals, enzymes, and lubricants necessary for the functioning of the digestive system.

Which cells of the **pancreas** secrete **enzymes**?

The pancreas, which also functions as an endocrine gland as described in the chapter on the endocrine system, consists of both endocrine and exocrine cells. The acinar cells (also called acini, from the Latin meaning "grapes" because their structure resembles clusters of grapes) are responsible for secreting digestive enzymes.

265

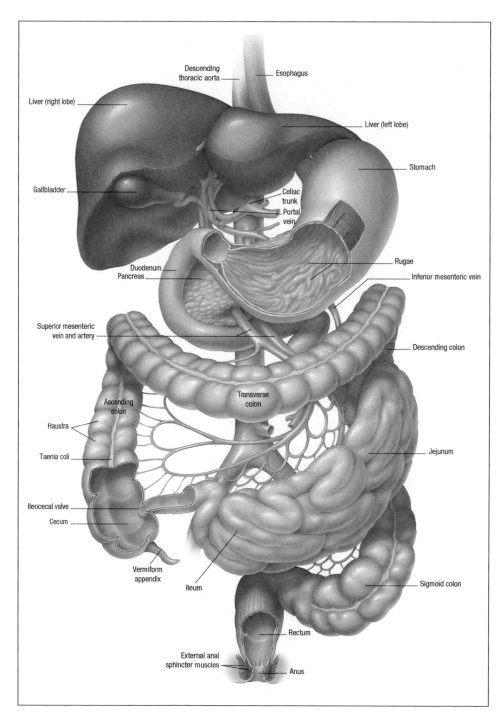

The digestive system includes not only the stomach and intestines, but also accessory organs such as the liver and gall bladder. *Anatomical Chart Co*.

How do the **pancreatic enzymes** reach the **small intestine**?

The pancreatic enzymes reach the small intestine via the heptatopancreatic duct. This duct is formed by linking the bile duct with the pancreatic duct. The pancreatic secretions are highly alkaline in order to neutralize the acidic chyme.

How much **pancreatic digestive juice** is secreted daily?

Nearly 1.6 quarts (1.5 liters) of digestive juices are secreted by the cells of the pancreas daily.

What is the **importance** of **pancreatic digestive juices**?

The pancreatic digestive juices are an alkaline solution (pH 8) composed of many enzymes. These enzymes are able to breakdown all categories of food.

How **large** is the **liver**?

The liver is the second largest organ in the body (the skin is the largest). It weighs 3 pounds (1.4 kilograms) in adults, representing about 2.5 percent of the total body weight. In children, the liver accounts for 4 percent of total body weight. The characteristic pudgy abdomen of infants is a result of the size of the liver.

What are the **digestive functions** of the **liver**?

The liver has more than 500 vital functions. Its major function as a digestive organ is to produce and secrete bile. Other functions of the liver include separating and filtering waste products from nutrients, storing glucose, and producing many chemical substances, such as cholesterol and albumin.

What is **bile**?

Bile is an alkaline liquid composed mostly of water, bile salts, bile pigments (bilirubin), fats, and cholesterol. It is essential for digestion of fats because it breaks down fats into fatty acids, which can then be absorbed by the digestive tract.

Why does **bile** have a **yellow-green** color?

Bile gets its color from bilirubin. Bilirubin is a waste product from the breakdown of worn-out red blood cells.

How does alcohol affect the liver?

Excessive alcohol use damages the liver. Since the liver is the chief organ responsible for metabolizing alcohol, it is especially vulnerable to alcohol-related injury. Alcoholic liver disease includes three conditions: 1) fatty liver (steatosis), 2) alcoholic hepatitis (inflammation of the liver), and 3) cirrhosis. The only effective treatment is to stop drinking.

What are some causes of **cirrhosis** of the liver?

The most common causes of cirrhosis in the United States are alcoholism and hepatitis C. Worldwide, hepatitis B is probably the most common cause of cirrhosis. In cirrhosis of the liver, scar tissue replaces normal, healthy tissue, blocking the flow of blood through the organ and preventing the liver from working as it should. Although liver damage from cirrhosis cannot be reversed, treatment can stop or delay further progression and reduce complications.

What is **hepatitis**?

Hepatitis is inflammation of the liver and commonly results from a viral infection. Hepatitis may either be acute or chronic. Acute hepatitis is short-lived, while chronic hepatitis is an inflammation of the liver that lasts for at least six months. Symptoms of acute hepatitis usually begin suddenly and include loss of appetite, nausea, vomiting, abdominal pain, low-grade fever, fatigue, and jaundice. Jaundice (from the Old French *jaune,* meaning "yellow") is a yellow discoloration of the skin and the whites of the eyes. Many individuals with chronic hepatitis have mild symptoms. Depending on the type of viral hepatitis, it may resolve itself on its own without medical intervention.

What are the main **causes** of **nonviral hepatitis**?

The two main types of nonviral hepatitis are alcoholic hepatitis and toxin/drug-induced hepatitis. Alcoholic hepatitis is the result of excessive drinking and often leads to cirrhosis. The inhalation or ingestion of certain toxins, such as carbon tetrachloride, vinyl chloride, and poisonous mushrooms are causes of hepatitis. Certain drugs including large dosages of the pain reliever acetaminophen may cause hepatitis.

Which types of **hepatitis** are **risk factors** for **liver cancer**?

According to the National Cancer Institute, the most important risk factor for liver cancer is a chronic infection with hepatitis B or hepatitis C.

How many different **types** of **viral hepatitis** have been identified?

There are five main types of viral hepatitis: hepatitis A, hepatitis B, hepatitis C, hepatitis D, and hepatitis E.

Comparison of Hepatitis A–E

Type of Hepatitis	Transmission	Treatment	Prevention
Hepatitis A	Food or water contaminated by feces from an infected person due to poor hygiene	Usually resolves itself without medical intervention within a few weeks	Practice good hygiene and sanitation; hepatitis A vaccine; avoid contaminated water (even tap water) in areas where the disease is widespread
Hepatitis B	Contact with infected blood; sexual relations (both heterosexual and homosexual) with an infected person; from mother to child during birth	Acute hepatitis B usually resolves itself without medical intervention; chronic hepatitis B may be treated with medications	Hepatitis B vaccine; avoid high-risk behaviors such as sharing needles and multiple sexual partners
Hepatitis C	Contact with infected blood (e.g. sharing needles); rarely through sexual contact and childbirth	Acute hepatitis C may resolve itself within a few months; chronic hepatitis C is treated with medications	Avoid high-risk behaviors such as sharing needles; avoid sharing personal items, such as razors and toothbrushes, with infected individuals; there is no vaccine
Hepatitis D	Contact with infected blood; only occurs in people who are already infected with hepatitis B	Chronic hepatitis D is treated with the drug alpha interferon	Avoid exposure to infected blood, contaminated needles, and an infected person's personal items (toothbrush, razor, nail clippers); immunization against hepatitis B
Hepatitis E	Food or water contaminated by feces from an infected person; uncommon in the United States	Usually resolves itself within several weeks or months	Reduce exposure to the virus by avoiding tap water when traveling internationally and practicing good hygiene and sanitation; there is no vaccine

What is the major **purpose** of the **gall bladder**?

The gall bladder (from the Latin *galbinus,* meaning "greenish yellow"), a pear-shaped, small sac, is mainly a storage vessel. It is connected by ducts to both the liver and small intestine. It stores bile until it is needed in the duodenum. Its name is derived from its usual color of green from the accumulation of bile.

What are the **two types** of **gallstones**?

Gallstones are hardened masses (stones) of bile. Gallstones form when bile contains too much cholesterol, bile salts, or bilirubin. The two types of gallstones are cholesterol stones and pigment stones. Cholesterol stones are more common, accounting for nearly 80 percent of all instances of gallstones. They are usually yellow-green in color and are made primarily of hardened cholesterol. An insufficient amount of water may also contribute to the development of cholesterol gallstones. Pigment stones are small, dark stones made of bilirubin.

How **serious** are **gallstones**?

Many individuals who have gallstones are asymptomatic and treatment is not necessary. However, if the stones block a duct, bile may be prevented from entering the small intestine. Surgery is then often recommended to remove the gall bladder. Women between 20 and 60 years of age are twice as likely to develop gallstones as men.

METABOLISM AND NUTRITION

What is **metabolism**?

Metabolism (from the Greek *metabole,* meaning "change") refers to the physical and chemical processes involved in the activities of the body. It includes the conversion of nutrients into usable energy contained in ATP, the production and replication of nucleic acids, the synthesis of proteins, the physical construction of cells and cell parts, the elimination of cellular wastes, and the production of heat, which helps regulate the temperature of the body.

How do **catabolic** reactions differ from **anabolic** reactions?

Catabolic and anabolic reactions are metabolic processes. A catabolic reaction is one that breaks down large molecules to produce energy; an example is digestion. An anabolic reaction is one that involves creating large molecules out of smaller molecules; an example is when your body makes fat out of extra nutrients you eat.

What are the **essential nutrients**?

There are six essential nutrients: carbohydrates, fats, proteins, water, vitamins, and minerals. The chemistry and basic biology that describes carbohydrates, fats, proteins and water are discussed in chapter 2 (Basic Biology).

Which nutrients are **energy nutrients**?

Energy nutrients are those that provide the body with the majority of the energy needed for daily metabolic reactions. Carbohydrates, fats, and proteins are energy nutrients.

What is a **calorie**?

A calorie is the amount of energy (heat) required to raise 1 gram (1 milliliter) of water by 1°C. A kilocalorie (kcal) is the amount of energy required to raise 1 kilogram (1 liter) of water by 1°C. The kilocalorie is the unit used to describe the energy value in food, since the calorie is a relatively small unit of measurement. For example, if a chocolate chip cookie were completely incinerated, the amount of heat energy released would be enough to raise the temperature of one liter of water by approximately 300°C.

What is the **comparative value** of common **energy sources** for cells?

The table below shows comparative values of energy sources.

Energy Source	Energy Yield (kcal/g)
Carbohydrate	4
Fat	9
Protein	4

What are **vitamins** and **minerals**?

A vitamin is an organic, nonprotein substance that is required by an organism for normal metabolic function but that cannot be synthesized by that organism. In other words, vitamins are crucial molecules that must be acquired from outside sources. While most vitamins are present in food, vitamin D, for example, is produced as a precursor in our skin and converted to the active form by sunlight. Minerals, such as calcium and iron, are inorganic substances that also enhance cell metabolism. Vitamins may be fat- or water-soluble.

Functions and Sources of Vitamins

Vitamin	Fat-/Water-Soluble	Major Sources	Major Functions
A	Fat-soluble	Animal products; plants contain only vitamin A building blocks	Aids normal cell division and development; particularly helpful in the maintenance of visual health

Vitamin	Fat-/Water-Soluble	Major Sources	Major Functions
B-complex	Water-soluble	Fruits and vegetables (folate); meat (thiamine, niacin, vitamin B_6, and B_{12}); milk (riboflavin, B_{12})	Energy metabolism; promotes harvesting energy from food
C	Water-soluble	Fruits and vegetables, particularly citrus, strawberries, spinach, and broccoli	Collagen synthesis; antioxidant benefits; promotes resistance to infection
D	Fat-soluble	Egg yolks; liver; fatty fish; sunlight	Supports bone growth; maintenance of muscular structure and digestive function
E	Fat-soluble	Vegetable oils; spinach; avocado; shrimp; cashews	Antioxidant
K	Fat-soluble	Leafy, green vegetables; cabbage	Blood clotting

Which **diseases** are caused by **vitamin deficiencies**?

Certain diseases are linked to dietary deficiencies of certain vitamins. Insufficient quantities of vitamin C in the diet leads to scurvy. Pellagra results from a lack of niacin (vitamin B_3) in the diet. Rickets in children and the related disease of osteomalacia in adults is caused by a lack of vitamin D.

Is it possible to have an **overabundance** of **vitamins**?

Excessive intake of vitamins may cause health complications as serious as vitamin deficiencies. The clinical term for excessive intake of vitamins is hypervitaminosis. It occurs when the dietary intake of a vitamin exceeds the body's ability to store, utilize, or excrete the vitamin. Hypervitaminosis is most common among the fat-soluble vitamins because the excessive quantities of the vitamins may be stored in lipid tissues. Water-soluble vitamins do not accumulate in the body since they are excreted in urine.

Which **vitamin** is **most frequently** consumed in **excess** amounts?

Vitamin A toxicity caused by massive dosages (from ten to thousands of times the recommended daily allowance) is the most common form of hypervitaminosis. Symptoms of vitamin A overdose include nausea, vomiting, headache, dizziness, and lethargy. Chronic overdose can lead to hair loss, joint pain, hypertension, weight loss, liver enlargement, and possibly death.

Are **vitamin supplements** necessary?

Vitamin supplements may be a useful addition to the diet of individuals who do not receive all of the nutrients they need from their diet. These individuals cannot or do not eat enough, or do not eat enough of a variety of healthy foods.

What is the **Food Guidance System**?

The Food Guidance System (MyPyramid) is an educational tool to help individuals implement the *Dietary Guidelines for Americans* published jointly by the Department of Health and Human Services (HHS) and the U.S. Department of Agriculture (USDA). The current system maintains the shape of the familiar food guide

The Food Pyramid, which was revised by the U.S. Department of Agriculture in 2005, is divided into colored sections. From left to right, the sections are: grains, vegetables, fruits, fat and oils, milk, meat and beans. *USDA Center for Nutrition Policy and Promotion.*

pyramid, but it is personalized for age, sex, and individual physical activity levels. The new pyramid symbol in MyPyramid features six vertical color bands representing the five food groups and oils. Each food group narrows toward the top to indicate moderation. The stylized symbol of a person climbing the steps is to encourage physical activity.

How do **saturated fats** differ from **unsaturated fats**?

Saturated fats and unsaturated fats are dietary fats. Saturated fats do not have double bonds between their carbon atoms and are solids at room temperature. Unsaturated fats have one or more double bonds between their carbon atoms and are liquids at room temperature. They are considered to be oils. Polyunsaturated fats have many double bonds.

What are **trans fatty acids**?

Trans fatty acids, or trans fats, are made when manufacturers add hydrogen to liquid vegetable oil—a process called hydrogenation—creating solid fats like shortening and hard margarine. Hydrogenation increases the shelf life and flavor stability of foods containing these fats. Diets high in trans fat raise the LDL (low density lipoprotein) or "bad" cholesterol, increasing the risk for coronary heart disease.

Which **foods** contain **trans fats**?

Cakes, crackers, cookies, snack foods, and other foods made with or fried in partially hydrogenated oils are the largest source (40 percent) of trans fats in the

American diet. Animal products and margarine are also major sources of trans fats. Since January 2006, the U.S. government has directed that the amount of trans fat in a product must be included in the Nutrition Facts panel on food labels.

What are **eating disorders**?

Eating disorders are medical illnesses in which patients become obsessed with food and their body weight. Research indicates that more than 90 percent of those who have eating disorders are women between the ages of 12 and 25. The main types of eating disorders are anorexia nervosa and bulimia nervosa. A third disorder, binge eating disorder, is still being investigated by researchers.

What is **anorexia**?

Anorexia simply means a loss of appetite. Anorexia nervosa is a psychological disturbance that is characterized by an intense fear of being fat. This persistent "fat image," however untrue in reality, leads the patient to self-imposed starvation and emaciation to the point where one-third of the body weight is lost. Clinical diagnosis of anorexia is determined when patients weigh at least 15 percent less than the normal healthy weight for their height. Many patients do not maintain a normal weight because they refuse to eat enough or avoid eating, exercise obsessively, induce vomiting, and/or use laxatives or diuretics to lose weight. For women, this causes the menstrual period to stop.

What is **bulimia**?

Bulimia is an eating disorder in which individuals binge eat frequently—often several times a week or even several times per day. Sufferers of this illness may eat an enormous amount of food in a short time, consuming thousands of calories. Then they will purge their bodies by vomiting or using laxatives and/or diuretics.

How does **binge eating disorder** differ from bulimia?

Individuals with binge eating disorder also consume large quantities of food in a short period of time. However, unlike individuals with bulimia, there is no purging.

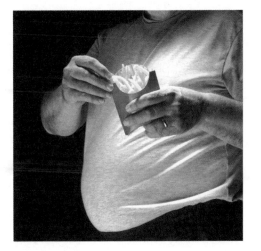

Fast foods are high in fats, so doctors generally tell their patients to avoid eating at fast food chains. Poor diet and lack of exercise are the main reasons for obesity. © iStockphoto.com/Atonal Arts.

What are some **causes** of **nausea** and **vomiting**?

Nausea, the sensation of having the urge to vomit, and vomiting, the emptying of the contents of the stomach through the mouth, may be caused by a variety of different reasons. Some common causes of nausea and vomiting are:

- Gastroenteritis, commonly known as "stomach flu," from bacterial or viral infections
- Food poisoning
- Overeating
- Migraine headaches
- Brain injury or concussions
- Inner ear and balance disorders
- Motion sickness
- Hormonal imbalances, especially during the first trimester of pregnancy
- Certain toxins, such as alcohol
- Drugs, such as chemotherapy agents, for treating cancer

How does the **definition** of "**overweight**" differ from "**obesity**"?

Both "overweight" and "obesity" describe ranges of weight that are greater than what is generally considered healthy for a given height. An objective measure of these terms is based on body mass index. An adult with a body mass index of 25 to 29.9 is considered overweight. An adult with a body mass index greater than 30 is considered obese.

What are the **causes** of **obesity**?

Obesity is a result of an individual consuming more calories than she or he burns. Genetic, environmental, psychological, and underlying medical problems may all be factors that lead to obesity. Underlying medical conditions may include hypothyroidism, Cushing's syndrome, depression, and certain neurological problems that can lead to overeating. Drugs, such as steroids, used to treat certain medical conditions may cause weight gain, too. Scientific study has indicated that obesity is linked to heredity. However, since family groups also share the same basic diet, it is difficult to separate genetics from environment. The average American diet tends to include many foods that are high in fat. The standard portion size for many foods has increased over the past several years. "Super-sized" portions available at fast-food restaurants have more calories. This diet coupled with a lack of physical activity leads to obesity. In addition, many people overeat in response to negative emotions such as sadness, anger, and boredom.

How is **body mass index (BMI)** calculated?

Body mass index (BMI) is a number that shows body weight adjusted for height. The formula to calculate BMI is: BMI = (weight in pounds $\times$ 703)/(height in inches)^2.

Evaluating BMI in Adults

BMI	Weight Status
Below 18.5	Underweight
18.5–24.9	Normal
25.0–29.9	Overweight
30.0 and above	Obese

The BMI for children and teens (ages 2 to 20) is age- and gender-specific. Healthcare professionals use established percentile cutoff points to identify underweight and overweight children. The percentile indicates the relative position of the child's BMI relative to children of the same sex and age.

Evaluating BMI in Children and Teenagers Ages 2 to 20

BMI	Weight Status
BMI-for-age < 5th percentile	Underweight
BMI-for-age 5th percentile to < 85th percentile	Normal
BMI-for-age 85th percentile to < 95th percentile	At risk of being overweight
BMI-for-age $\geq$ 95th percentile	Overweight

What **health risks** are associated with **obesity**?

People who are obese are more likely to develop a variety of health problems, including:

- Hypertension
- Dyslipidemia (for example, high total cholesterol or high levels of triglycerides)
- Type 2 diabetes
- Coronary heart disease

- Stroke
- Gall bladder disease
- Osteoarthritis
- Sleep apnea and respiratory problems
- Some cancers (endometrial, breast, and colon)

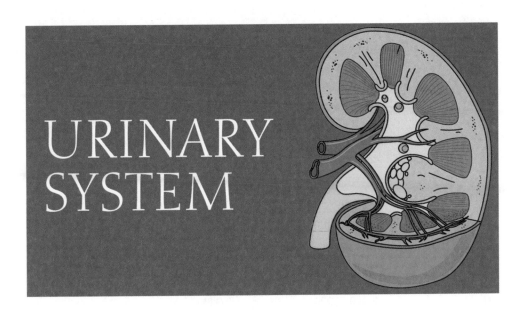

URINARY SYSTEM

INTRODUCTION

What are the **functions** of the **urinary system**?

The functions of the urinary system include regulation of body fluids, removal of metabolic waste products, regulation of volume and chemical make-up of blood plasma, and excretion of toxins.

What are the **major parts** of the **urinary system**?

The major parts of the urinary system are the kidneys, the urinary bladder, two ureters, and the urethra. Each component of the urinary system has a unique function. Urine is manufactured in the kidneys. The urinary bladder serves as a temporary storage reservoir for urine. The ureters transport urine from the kidney to the bladder, while the urethra transports urine from the bladder to the outside of the body.

What are the **sources** of **water gain** and **loss** per day?

The main sources of water gain are drinking and ingesting fluids, such as water contained in food, and water produced as a byproduct of metabolic processes. The main sources of water loss are urine formation, evaporation from the lungs (breathing), evaporation from the skin (sweating), and through the feces.

Sources of Water Gain and Loss per Day

Fluid Intake Source	mL	Fluid Output Source	mL
Ingested liquid	1,200–1,500	Urine	1,200–1,700
Ingested food	700–1,000	Feces	100–250

Fluid Intake Source	mL	Fluid Output Source	mL
Metabolic oxidation	200–400	Sweat	100–150
		Insensible losses	
		Skin	350–400
		Lungs	350–400
TOTAL	2,100–2,900	TOTAL	2,100–2,900

What **percent** of a person's intake of **water** comes from **drinking water**?

Only about 47 percent of a person's daily water intake comes from drinking. Nearly 39 percent of water intake comes from eating solid food, since water is a major component of many foods. For example, fruits and vegetables may contain more than 90 percent water.

What is **cosmetic dehydration**?

Cosmetic dehydration is the practice of taking large doses of diuretics to cause temporary weight loss. It has been used by fashion models and body builders, but it is a dangerous practice because it can cause electrolyte imbalance and cardiac arrest.

KIDNEYS

How **big** is a human **kidney**?

The human kidney is about the size of a human fist and weighs about 5 ounces (150 grams). Its average dimensions are 5 inches (12 centimeters) long, 3 inches (6 centimeters) wide, and 1 inch (2.5 centimeters) thick.

Where are the **kidneys located**?

The kidneys are located on either side of the spinal column in the lumbar region, just underneath the ribcage. Their exact anatomic position is retroperitoneal (between) the dorsal body wall and the parietal peritoneum.

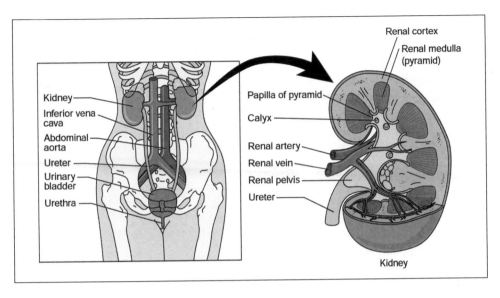

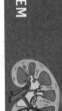

The kidney and urinary tract. (From Smeltzer, S. C., and Bare, B. G. *Textbook of Medical-Surgical Nursing*. 9th Ed. Philadelphia: Lippincott, Williams & Wilkins, 2000.)

What are the **parts** of the **kidney**?

The kidney has two layers: the outer layer, called the cortex, which is reddish brown and granular, and the inner zone, the medulla, which is darker and reddish brown in color. The medulla is subdivided into 6 to 18 cone-shaped sections called the pyramids. The pyramids are inverted so that each base faces the cortex and the tops project toward the center of the kidney. Separating the pyramids are bands of tissue called renal columns. A renal lobe consists of a renal pyramid and its surrounding tissue.

How are the **kidneys protected** in the abdominal cavity?

The kidneys receive some protection from the lower part of the rib cage. In addition, three layers of connective tissue enclose, protect, and stabilize the kidneys: 1) the renal capsule that covers the outer surface of the entire organ; 2) the adipose (fatty) tissue that surrounds the capsule; and 3) the renal fascia, which is a layer of dense connective tissue that anchors the kidney to the surrounding structure.

Why is it **important** to **protect** the **kidneys**?

The kidneys are protected from day-to-day jolts and shocks by the adipose tissue and renal fascia. If the fibers of the dense connective tissue break, the kidney will be able to move to the abdominal area. Movement of the kidney, called floating kidney or nephroptosis, is dangerous because the ureters of renal blood vessels may become twisted.

What **gland** is found on **top** of the **kidneys**?

The adrenal glands are closely associated with the kidneys, with one adrenal gland sitting on top of each kidney and embedded in fatty tissue that encloses the kidney.

How do **kidneys** help **vitamin D** be available for bone growth?

The kidneys turn vitamin D into an active hormone called calcitrol, which helps bones absorb the right amount of calcium from blood. If the kidneys are impaired, bones do not get enough calcium, either because the kidneys fail to turn vitamin D into calcitrol or because they allow too much phosphorus to build up in the blood. The excess phosphorus draws calcium into the blood and blocks it from getting to the bones.

What is a **nephron**?

A nephron is the functional working unit of the kidney. Blood is filtered in the nephrons and toxic wastes are removed, while water and necessary nutrients are reabsorbed into the system. Each nephron produces a minute amount of urine, which then trickles into the renal pelvis. From there it goes into the ureter and eventually collects in the bladder.

How many **nephrons** are there in **one kidney**?

There are approximately one million nephrons in each kidney.

What are the **major parts** of the **nephron**?

The major parts of the nephron are the renal corpuscle, the proximal convoluted tubule, the loop of Henle, and the distal convoluted tubule. Each part has a significant, distinct function in urine production. Inside the renal corpuscle is a tangled network of about 50 capillaries, the glomerulus, where filtration occurs. The blood pressure within the capillaries forces water and dissolved substances out of the capillaries and into the renal tubule.

The proximal convoluted tubule starts at the glomerulus and continues into the loop of Henle. About 65 percent of the original filtrate is reabsorbed in the proximal convoluted tubule. The loop of Henle is a critical region where the filtrate is further adjusted for water and solute balance. The distal convoluted tubule removes additional water from the filtrate, until the final concentration of the urine is approximately equal to that of the body fluids.

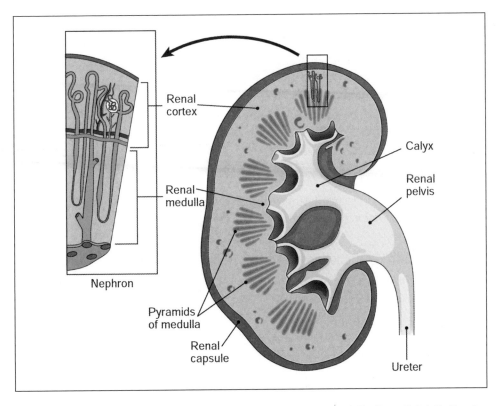

Renal cortex

Calyx

Renal pelvis

Renal medulla

Nephron

Pyramids of medulla

Renal capsule

Ureter

The nephron is the functioning unit of the kidney. (Cohen, B. J., and Wood, D.L. *Memmler's The Human Body in Health and Disease*. 9th Ed. Philadelphia: Lippincott, Williams & Wilkins, 2000.)

What is the **vasa recta**?

The vasa recta are the long, straight blood vessels that extend into the medulla of the kidney and parallel to the course of the loop of Henle. There is a distinct difference in solute concentration between the fluid in the loop of Henle and the plasma concentration in the vessels of the vasa recta. This mechanism is called counter current exchange; it results in the reabsorption of water and solutes between the fluid in the loop of Henle and blood.

What is **urea** and where is it produced?

During the process of metabolizing proteins, the body produces ammonia. Ammonia combines with carbon dioxide to form urea. The human body can tolerate 100,000 times more urea than ammonia. It is the most abundant organic waste produced in the body and it is eliminated by the kidneys. Humans generate about 0.75 ounces (21 grams) of urea each day.

What are the **major vessels** that enter and leave the **kidneys**?

Each kidney receives blood from a renal artery. The kidneys receive 20 to 25 percent of the total cardiac output, or approximately 2.5 pints (1,200 milliliters) of blood per minute. After circulating through the kidney and nephrons, the blood is collected in the renal vein. The renal nerves innervate the kidneys.

How much **fluid** do the kidneys **remove** from the blood **daily**?

The kidneys filter about 48 gallons (182 liters) of blood daily and produce about 4 ounces of filtrate per minute. About 1.5 to 2 quarts (1.4 to 1.9 liters) of urine is eventually excreted per day.

How much **blood** is **filtered** by the kidneys in an average **lifetime**?

The entire blood supply is filtered through the kidneys 60 times per day. The kidneys in a person living 73 years filter almost 1.3 million gallons of blood.

How do the **kidneys** help control **red blood** cell **production**?

The kidneys produce erythropoietin, a hormone that affects the number of circulating red blood cells.

What is a **kidney stone**?

Kidney stones, or renal calculi, are the precipitates of substances such as uric acid, calcium oxalate, calcium phosphate, and magnesium phosphate that usually form in the renal pelvis. A stone passing into a ureter can cause very severe pain. Approximately 50 percent of kidney stones pass from the body on their own. Stones were once removed surgically, but most are now shattered with sound waves in a procedure called lithotripsy. Stones may form in the ureter or bladder, in addition to the kidneys.

What is the name of the **process** when **stones** are formed in the **urinary tract**?

The process of stone formation is called urolithiasis, renal lithiasis, or nephrolithiasis.

What are the **symptoms** of **kidney failure**?

Symptoms of kidney failure include excess fluid buildup due to inability of the kidneys to produce enough urine. The resulting excess fluid increases blood pressure,

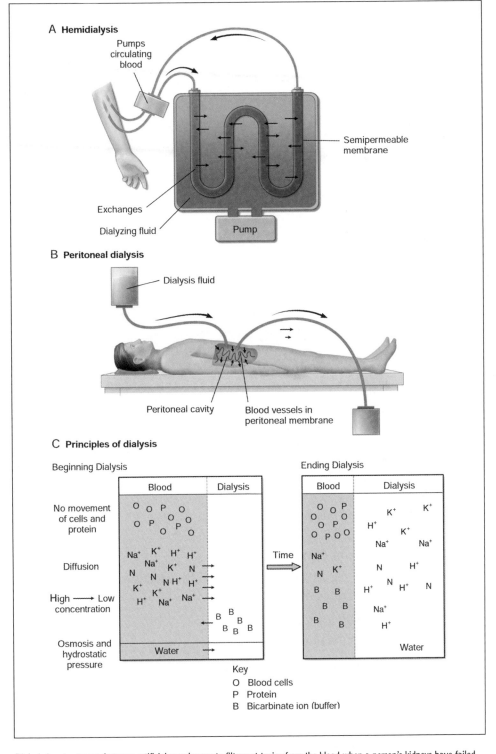

A Hemidialysis

Pumps circulating blood

Semipermeable membrane

Exchanges

Dialyzing fluid

Pump

B Peritoneal dialysis

Dialysis fluid

Peritoneal cavity

Blood vessels in peritoneal membrane

C Principles of dialysis

Beginning Dialysis

Ending Dialysis

	Blood	Dialysis
No movement of cells and protein	O O P O O P O O O P O	
Diffusion	Na⁺ K⁺ H⁺ H⁺ N Na⁺ K⁺ N N N H⁺ H⁺ K⁺ N K⁺ Na⁺ Na⁺	
High ⟶ Low concentration	H⁺ Na⁺ Na⁺	B B B B B
Osmosis and hydrostatic pressure	Water ⟶	

Time

	Blood	Dialysis
	O O P O O P O O P O O O P O O	K⁺ K⁺ H⁺ K⁺ Na⁺ Na⁺
	Na⁺ N K⁺ B B B B	N H⁺ H⁺ N H⁺ N Na⁺ H⁺
		Water

Key

O   Blood cells
P   Protein
B   Bicarbinate ion (buffer)

Dialysis is a treatment that uses artificial membranes to filter out toxins from the blood when a person's kidneys have failed. (From Premkumar K. *The Massage Connection Anatomy and Physiology.* Baltimore: Lippincott, Williams & Wilkins, 2004.)

leading to hypertension. An indirect symptom of kidney failure is anemia due to the decreased production of erythropoietin by the kidneys. Erythropoietin controls the rate of maturation of red blood cells. Without adequate red blood cells, a person will become tired and short of breath.

What conditions can cause **chronic renal disease**?

Diabetes mellitus is the leading cause of chronic renal disease. It accounts for about 44 percent of new cases each year. The second leading cause is hypertension (high blood pressure).

What is kidney **dialysis**?

Kidney dialysis, also known as hemodialysis, is used when the kidneys are not functioning or are improperly functioning. An artificial membrane is used to regulate blood composition, particularly the removal of toxic substances. The patient's blood flows through artificial membrane tubing, which is immersed in a solution that differs in concentration from the normal concentration of blood plasma. Critical to this process is the composition of the dialysis fluid, which permits retention of needed substances while wastes are removed. Dialysis is usually carried out three times per week.

How long does it take to complete **dialysis**?

The procedure usually takes about four hours, but the exact time needed for dialysis is dependent on the following:

- How well the kidneys are functioning
- The amount of fluid retained
- The amount of waste present in the blood
- Body weight
- Type of artificial kidney used

Does **dialysis cure** kidney disease?

Dialysis does not cure kidney disease. It does the work of your kidneys, but it cannot replace them permanently.

Which human **organ** was the **first** to be **transplanted**?

The first human organ to be successfully transplanted was the kidney. Dr. Joseph Murray (1919–) performed the transplant in 1954 in Boston, Massachusetts. The patient, Richard Herrick, lived for eight years after receiving the new kidney from his identical twin brother, Ronald Herrick.

How **successful** are kidney **transplants**?

The one-year success rate for kidney transplants is 85 to 95 percent.

What are the **risks** associated with a **kidney transplant**?

As with any transplant, rejection of the foreign body is the major cause of transplant failure. Recipients of kidney transplants have to take immunosuppressants for the rest of their lives.

ACCESSORY ORGANS

How long is each **ureter**?

Each ureter is 10 to 12 inches (25 to 30 centimeters) long. The ureters extend from the kidney into the bladder. They begin as thin, hollow, narrow tubes and widen to 0.5 inches (1.7 centimeters) as they enter the bladder. Urine is transported to the urinary bladder via the two ureters.

Where is the urinary **bladder located**?

The urinary bladder is located in the abdominal cavity. In males, it is anterior to the rectum and above the prostate gland. In females, it is located much lower, anterior to the uterus and upper vagina.

How much **urine** can the urinary **bladder hold**?

The urinary bladder is highly distensible and can vary in its capacity. As urine fills the bladder, it can expand to about 5 inches (12 centimeters) long and hold 1 pint (473 milliliters) of urine at moderate capacity. The bladder can expand to twice that capacity if necessary. It usually accumulates 300 to 400 milliliters of urine before emptying, but it can expand to hold 600 to 800 milliliters.

Why do **pregnant women** have an **increased** need and **urge** to **urinate**?

Early in a pregnancy, hormonal changes in the mother make the urge to urinate more frequent and more urgent. Towards the end of a pregnancy, the increased size and weight of the uterus pressing on the bladder reduces its capacity to hold urine, causing an increased need to urinate.

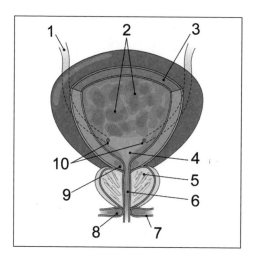

The bladder and surrounding structures: 1 = Ureter; 2 = Rugae; 3 = Smooth muscle; 4 = Trigone; 5 = Prostate; 6 = Urethra; 7 = Urogenital diaphragm; 8 = External sphincter; 9 = Internal sphincter; 10 = Openings of ureters. (Cohen, B. J., and Wood, D. L. *Memmler's The Human Body in Health and Disease*. 9th Ed. Philadelphia: Lippincott, Williams & Wilkins, 2000.)

How does the **urethra** differ in **males** and **females**?

Urine is transported to the outside through the urethra, which is a thin, muscular tube that extends from the urinary bladder to the exterior of the body. The length and structure of the urethra differs between males and females. In males, the urethra is about 8 inches (20 centimeters) long and extends from the urinary bladder to the exterior. It has the dual function of transporting semen as well as urine out of the body. The female urethra is only about 1.5 inches (3 to 4 centimeters) long and extends from the bladder to the exterior opening.

Why are **women** more prone to **urinary tract infections**?

Women are more likely to suffer from urinary tract infections because the urethra in women is much shorter. It is easier for bacteria to reach the bladder, causing an infection.

URINE AND ITS FORMATION

What is the **composition of urine**?

Urine is mostly water and contains organic waste products such as urea, uric acid, and creatinine. It also contains excess ions, such as sodium (Na^+), potassium (K^+), chloride (Cl^-), bicarbonate (HCO_3^-), and hydrogen (H^+).

Is **urine** always **yellow** in color?

Normally, dilute urine is nearly colorless. Concentrated urine is a deep yellow; colors other than yellow are not normal. Food pigments can make the urine red, and drugs can produce colors such as brown, black, blue, green, or red. Urine may also be brown, black, or red due to disorders or diseases such as severe muscle injury or melanoma. Cloudy urine suggest the presence of pus, due to a urinary tract infection, or salt crystals from uric acid or phosphoric acid.

What is a **urinalysis**?

Urinalysis is the chemical and physical analysis of a urine sample. It involves the color and appearance of urine, plus a detailed list of specific compounds and their

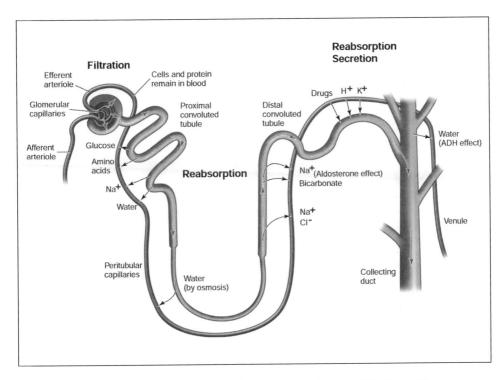

The process of urine formation. (From Premkumar K. *The Massage Connection Anatomy and Physiology*. Baltimore: Lippincott, Williams & Wilkins, 2004.)

concentration found in the sample. Substances that should not be found in urine are proteins, glucose, acetone, blood, and pus. The presence of any of these substances may indicate a disease.

Why is **urine used** to test for **drug use**?

Urine drug testing is commonly used for opioids and illicit drugs. The liver is where drug detoxification occurs, but the byproducts of this process are excreted by the kidneys. There are two types of tests: a screening test and a confirmatory test. Most urine drug tests screen for marijuana, cocaine, opiates, PCP, and amphetamines.

What are the main **parts** of **urine production**?

The main parts of urine production are glomerular filtration, tubular reabsorption, and tubular secretion. In filtration, blood pressure in the glomerular capillaries forces water and other solutes across the glomerular membrane. The process is similar to what happens in a coffee maker, where water passes through a filter and it carries with it some dissolved compounds (coffee). Usually, the coffee grinds never reach the pot, unless there is a hole in the filter. Reabsorption involves the return of water and major nutrients to the blood. Secretion is the removal of harmful or excess substances from the blood and their transport into the urine.

What is a **diuretic**?

A diuretic is a chemical that increases urine output. Examples of diuretics include alcohol and any beverages that contain caffeine (coffee, tea, colas). Diuretics are usually prescribed for patients with high blood pressure or congestive heart failure.

What is the **antidiuretic hormone (ADH)**?

The antidiuretic hormone (ADH) is one of the major hormones controlling fluid balance by inhibiting diuresis or urine output. ADH is secreted by the posterior pituitary gland. For example, when a person is dehydrated, the level of ADH secretion increases so that more water is reabsorbed from the urine and returned to the blood. Conversely, a diuretic such as alcohol inhibits the release of ADH and thus increases urinary output.

What is **micturition**?

Micturition, or urination, is the process that expels urine from the urinary bladder.

What **muscles** control **urination**?

The walls of the urinary bladder have three layers: 1) a mucosa layer made of transitional epithelium; 2) connective tissue; and 3) detrusor muscle, composed of smooth muscle fibers arranged in both longitudinal and circular layers. It is the contraction of the detrusor muscle that compresses the bladder when voiding urine.

What is **incontinence**?

Incontinence is the inability to control urination voluntarily. Contributing factors to incontinence include emotional problems, pregnancy, and nervous system problems.

REPRODUCTIVE SYSTEM

INTRODUCTION

What is the **function** of the **reproductive system**?

The function of the reproductive system is to produce new offspring. The continuity of the human species is guaranteed through reproduction.

What are the **organs** of the **reproductive system** and their general function?

The major organs of the reproductive system are the gonads, various ducts, accessory sex glands, and supporting structures. The gonads produce gametes and secrete hormones. The various ducts store and transport the gametes. The accessory sex glands produce substances that protect the gametes. The supporting structures assist the delivery and joining of gametes.

Are the **male** and **female** reproductive **systems identical**?

Unlike every other organ system in the human body, the male and female reproductive systems are not identical. The specialized organs of the male and female reproductive systems are different. However, it is essential for both the male and female reproductive systems to be involved in order for reproduction to occur.

Is the **reproductive system essential** for life?

The reproductive system is essential for the survival of the species. It is not essential for an individual to live and be healthy.

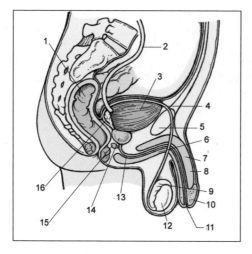 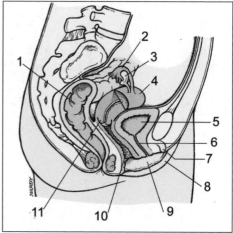

The male reproductive system and surrounding structures: 1 = Rectum; 2 = Ureter; 3 = Urinary bladder; 4 = Vas deferens; 5 = Pubic symphysis; 6 = Corpus spongiosum; 7 = Penis; 8 = Corpus cavernosum; 9 = Epididymus; 10 = Glans penis; 11 = Urethral orifice; 12 = Testis; 13 = Urethra; 14 = Prostate; 15 = Bulbourethral gland; 16 = Seminal vesicle. (From *Stedman's Medical Dictionary*. 27th Ed. Baltimore: Lippincott, Williams & Wilkins, 2000.)

The female reproductive system and surrounding structures: 1 = Rectum; 2 = Uterine tube; 3 = Ovary; 4 = Uterus; 5 = Urinary bladder; 6 = Urethra; 7 = Clitoris; 8 = Labia majora; 9 = Labia minora; 10 = Vagina; 11 = Cervix. (From *Stedman's Medical Dictionary*. 27th Ed. Baltimore: Lippincott, Williams & Wilkins, 2000.)

Which **branch of medicine** specializes in the reproductive system?

Gynecology (from the Greek *gune*, meaning "woman") is the branch of medicine that specializes in the diagnosis and treatment of conditions of the female reproductive system. Urologists treat medical conditions of the male reproductive system.

MALE REPRODUCTIVE SYSTEM

What are the **male reproductive organs** and structures?

The male reproductive organs and structures are the testes, a duct system that includes the epididymis and the vas deferens; the accessory glands, including the seminal vesicles and prostate gland; and the penis.

Which organs are the **male gonads**?

The testes are the male gonads. They produce the male reproductive cells called sperm.

Where are the **testes located**?

The testes hang in a pouch called the scrotum (from the Latin *scrautum*, meaning a "leather pouch for arrows"). The scrotum is a fleshy pouch consisting of loose skin, loose connective tissue, and smooth muscle. It is divided internally by a septum into two chambers each of which contains one testis.

Why is it advantageous for the **scrotum** to hang **outside the body**?

Since the scrotum hangs outside the body, the temperature in the scrotum is lower than normal body temperature. The normal temperature in the scrotum is 2–3°F (1.1–1.6°C) below normal body temperature. This lower temperature is necessary for production and survival of sperm.

How does the **scrotum react** to changes in **temperature**?

Changes in the external temperature stimulate the cremaster muscle to contract or relax, thereby moving the testes closer to or further from the body cavity. When exposed to cold temperatures, such as swimming in cold water, the cremaster muscle contracts, bringing the testes closer to the body to keep them warm. When temperatures are warm, the cremaster muscle relaxes, moving the testes farther away from the body so excess heat is lost via the increased surface area of the scrotum.

When do the **testes descend** into the scrotum during development?

The testes begin their development in the abdomen. They descend from the abdomen into the groin (the inguinal canal) during the third month of fetal development. The descent into the scrotum usually begins during the seventh month of fetal development and is completed shortly before or after birth.

Which **medical condition** occurs when one or both **testes do not descend** into the scrotum?

Failure of the testes to descend from the abdomen to the scrotum is called cryptorchidism (from the Greek *crypto,* meaning "hidden," and *orchis,* meaning "testis"). This condition occurs in three to five percent of male babies born at full-term. In premature babies the number of cases of cryptorchidism increases to 30 percent. In most cases, the undescended testicle(s) move into the scrotum without medical intervention. If necessary, the testicle(s) may be moved into the scrotum surgically. Testes are generally not left in the abdomen since this may interfere with their ability to produce sperm. Furthermore, testicular cancer is more common in men with undescended testicles.

Where does **sperm production** occur in the testes?

Sperm production, called spermatogenesis, occurs in the seminiferous tubules. There are approximately 800 seminiferous tubules in each testis. Each tubule is

293

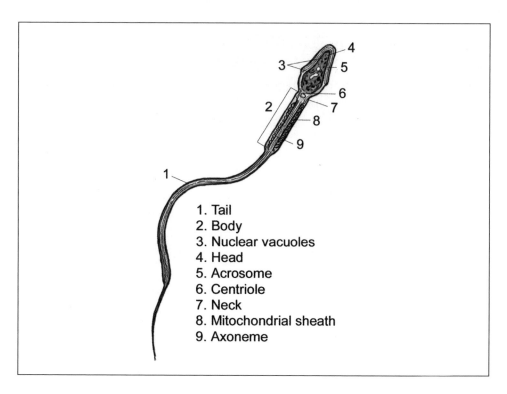

1. Tail
2. Body
3. Nuclear vacuoles
4. Head
5. Acrosome
6. Centriole
7. Neck
8. Mitochondrial sheath
9. Axoneme

The anatomy of a sperm cell. *Anatomical Chart Co.*

slender, tightly coiled, and approximately 31.5 inches (80 centimeters) long. Each testis contains nearly half a mile of seminiferous tubules. Spermatogenic (sperm-forming) cells line the seminiferous tubules.

What are the steps of **spermatogenesis**?

Spermatogenesis is the process whereby the seminiferous tubules produce sperm. There are three steps in spermatogenesis: 1) meiosis, during which the number of chromosomes in the cell is reduced to half or 23 chromosomes each; 2) meiosis II, during which each haploid cell forms spermatids; and 3) spermiogenesis, during which each spermatid develops into a sperm cell with a head and tail. The entire process of spermatogenesis takes about 64 days.

At what **age** does **sperm production** begin?

Sperm production begins with the onset of puberty, usually between ages 11 to 14 in boys. It continues throughout the life of an adult male.

How many sperm are produced?

It is estimated that during his lifetime a normal male will produce 10^{12} sperm, or as many as 300 million per day.

What are the **characteristics of sperm**?

Each sperm cell has three distinct regions: the head, the middle piece, and the tail. The head consists of a nucleus with the genetic material (DNA) and an acrosome at the tip. The acrosome contains enzymes to help the sperm penetrate the oocyte (egg). The middle piece contains mostly mitochondria which provide energy for movement. The tail moves the sperm cell from one place to another. The tail is the only flagellum in the human body.

How large are **sperm** cells?

Sperm are some of the smallest cells in the human body. A single sperm cell is 0.05 millimeters long from the head to the tip of the tail.

What are the **three ducts** through which **sperm travel** from the testes to the urethra?

When sperm leave the testes they have the physical characteristics of mature sperm, but they are not functionally mature. Final maturation occurs in the ducts, known as the accessory ducts. The three ducts are the epididymis, the ductus deferens, also known as the vas deferens, and the ejaculatory duct.

What are the **functions** of the **epididymis**?

The epididymis (from the Greek, meaning "upon the twin," with twin being the testis) is a long, twisted and coiled tubule. It occupies a space of about 1.5 inches (4 centimeters), but if uncoiled it would measure 20 to 23 feet (6 to 7 meters) long. While in the epididymis, sperm mature and gain motility. The epididymis stores sperm until they are ready to be ejaculated. The smooth muscle of the epididymis helps propel mature sperm into the ductus deferens via peristaltic contractions.

How **long** does it take **sperm to mature** in the **epididymis**?

Sperm spend 10 to 14 days in the epididymis prior to reaching final maturation. They may then be stored in the epididymis for up to a month. If they are not ejaculated during that time, they degenerate and are reabsorbed by the body.

What is the function of the **ductus deferens**?

The ductus deferens (from the Latin *deferre*, meaning "to carry away"), also called the vas deferens, carries sperm from the epididymis to the ejaculatory duct. It is only 16 to 18 inches (40 to 45 centimeters) long, much shorter than the epididymis, and it has a larger diameter.

Which is the **shortest duct** of the male reproductive system?

The ejaculatory duct is only about 1 inch (2.5 centimeters) long, much shorter than either the epididymis or the vas deferens. It is formed by the duct from the seminal vesicles and the vas deferens. The ejaculatory duct ejects sperm into the urethra.

What is the **reproductive function** of the **urethra**?

The urethra is the last section of the reproductive duct system. It provides a passageway to transport sperm outside of the body during ejaculation.

What are the **accessory glands** of the **male reproductive system**?

The accessory glands are the seminal vesicles, the prostate gland, and the bulbourethral glands. Each of these glands adds secretions to the sperm. These secretions constitute the liquid portion of semen.

Which **accessory gland** contributes the **greatest volume of secretions** for seminal fluid?

The seminal vesicles contribute nearly 60 percent of the volume of seminal fluid. These secretions are an alkaline fluid containing water, fructose, prostaglandins, and clotting proteins. The fructose provides an energy source for the sperm. Prostaglandins contribute to sperm motility and viability. The clotting proteins help semen coagulate after ejaculation. The alkalinity of the fluid helps to neutralize the acidic environments of the male urethra and female reproductive tract.

How large is the **prostate gland**?

The prostate gland in healthy adult males is about the size of a walnut. It is located in front of the rectum and under the bladder. The prostate surrounds the urethra. When it becomes enlarged, it squeezes the urethra, restricting the flow of urine.

What is the **purpose** of **prostate gland secretions**?

Prostate gland secretions account for nearly 30 percent of the volume of seminal fluid. The slightly acidic secretions help semen clot following ejaculation and then break down the clot.

What is **prostatitis**?

Prostatitis is an inflammation of the prostate gland. The two types of prostatitis are acute prostatitis and chronic prostatitis. It may be caused by bacterial pathogens or may develop without evidence of an infecting organism. Symptoms include pain in the lower back, difficulty with urination often accompanied by a burning sensation and/or pain, frequent urination, especially at night, and sometimes fever.

What **tests** are available to **screen** for **prostate cancer**?

The two tests available to screen for prostate cancer are the prostate-specific antigen (PSA) blood test and digital rectal exam (DRE). Prostate-specific antigen (PSA) is a protein produced by the cells of the prostate gland. The PSA test measures the level of PSA in the blood. A DRE involves a physician inserting a gloved, lubricated finger into the rec-

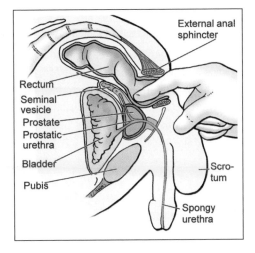

A simple prostate exam can help determine whether there is a swelling of the gland or possible cancerous tissue. (From Moore, K. L., Ph.D., FRSM, FIAC, and Dalley, A.F. II, Ph.D. *Clinical Oriented Anatomy.* 4th Ed. Baltimore: Lippincott, Williams & Wilkins, 1999.)

tum to feel the prostate for any irregularities or firm areas. The DRE is recommended in conjunction with the PSA to detect abnormalities in the prostate.

Are all **tumors** found in the **prostate malignant**?

No, benign prostatic hyperplasia is a condition an abnormal growth of benign (noncancerous, nonmalignant) cells in the prostate. Its most characteristic symptom is a reduced flow of urine. Most changes in the prostate are not cancer.

How **common** is **prostate cancer**?

Prostate cancer is the most common cancer, excluding skin cancers, in American men. It is the leading cause of cancer death among American men. The American Cancer Society estimated that 234,460 new cases of prostate cancer were diagnosed in the United States and 27,350 men in the United States died of prostate cancer during 2006. Approximately one man in every six will be diagnosed with prostate cancer during his lifetime, but only one man in 34 will die of this disease.

What is **semen**?

Semen (from the Latin, meaning "seed") is the combination of the secretions of the accessory glands and sperm. The composition of the seminal fluid is:

Secretions from the seminal vesicles: 60 percent
Secretions from the prostate: 30 percent
Secretions from the epididymis: 5 percent
Secretions from the bulbourethral glands: less than 5 percent
Sperm: about 1 percent.

How much **semen** is produced in a **typical ejaculation**?

An average, typical ejaculation contains 2 to 5 milliliters (approximately 0.5 to 1 teaspoon) of semen. There are 50 to 150 million sperm per milliliter in a typical ejaculation, or 300 to 400 million sperm in each ejaculation.

When is a man considered **sterile**?

Males with a sperm count below 20 million sperm per milliliter are considered functionally sterile, since too few spermatozoa survive to reach and fertilize an oocyte.

What is the **function** of the **penis**?

The penis is an external organ of the male reproductive system. It consists of three parts: 1) the root where the penis attaches to the wall of the abdomen; 2) the shaft or body; and 3) the glans, which is the tip of the penis. The body of the penis has a tubular, cylindrical shape and surrounds the urethra. Spongy tissue that can expand and contract fills the body or the penis. There is a slit opening at the tip of the glans through which urine is excreted and semen is ejaculated through the urethra.

What is **erectile tissue**?

The penis contains three cylindrical columns of erectile tissue; two are the corpus carvernosa and one is the corpus spongiosum. Upon stimulation, blood flow to the erectile tissue is increased resulting in an erection.

What are some causes of **erectile dysfunction**?

Erectile dysfunction (ED), the inability to achieve or sustain an erection, is often the result of disease, injury, or a side effect of certain drugs, including blood pressure drugs, antihistamines, antidepressants, tranquilizers, appetite suppressants, and cimetidine (an ulcer drug). Damage to nerves, arteries, smooth muscles, and fibrous tissues of the penis are the most common causes of erectile dysfunction. Diabetes, kidney disease, chronic alcoholism, multiple sclerosis, atherosclerosis, vascular disease, and neurologic disease account for about 70 percent of ED cases. Treatment may include lifestyle changes, adjusting medications to alleviate side effects, medications to induce erection, and surgery.

FEMALE REPRODUCTIVE SYSTEM

What are the **female reproductive organs**?

The organs of the female reproductive system include the ovaries, the uterine tubes, the uterus, the vagina, the external organs called the vulva, and the mammary glands.

An unfertilized ovum: 1 = Polar body; 2 = Nucleus; 3 = Nucleolus; 4 = Ooplasm; 5 = Zona pelucida; 6 = Corona radiata. *Anatomical Chart Co.*

Which organs are the **female gonads**?

The paired ovaries are the female gonads. Each ovary is approximately 1 to 2 inches (2.5 to 5.0 centimeter) long, 0.6 to 1.2 inches (1.5 to 3.0 centimeters) wide and 0.24 to 0.6 inches (0.6 to 1.5 centimeters) thick—similar in size and shape to an unshelled almond. They produce the female gametes, called ova, and secrete the female sex hormones.

What are the steps of **oogenesis**?

Oogenesis is the production of ova (eggs) in the ovaries. The process begins during fetal development. Meiosis I occurs prior to birth, and then development of the primary oocytes is suspended until puberty. The secondary oocyte is released at ovulation in response to hormonal secretions. If fertilization occurs, meiosis II resumes and an ovum is formed.

How many **oocytes** does a girl have **at birth**?

There are approximately two million oocytes in the ovaries at birth. Only about 300,000 to 400,000 of the oocytes survive until puberty, and only about 400 of the oocytes that survive until puberty will be released at ovulation during a woman's lifetime. The remainder of the oocytes degenerate.

Are the **uterine tubes connected** to the **ovaries**?

No, the superior, funnel-shaped end of the uterine tubes opens in the abdominal cavity very close to the ovaries, but is not actually connected to the ovaries. The inferior end of the uterine tubes opens into the uterus.

What event **occurs** in the **uterine tubes**?

Fertilization of the secondary oocyte by sperm usually occurs in the uterine tubes. It takes four to seven days following fertilization for a fertilized oocyte (zygote) to reach the uterus. Unfertilized oocytes degenerate.

What are the **functions** of the **uterus**?

The uterus (from the Latin, meaning "womb") is an inverted, pear-shaped structure located between the urinary bladder and the rectum. Implantation of a fertilized ovum occurs in the uterus. Following implantation, the uterus houses, nourishes, and protects the developing fetus during pregnancy.

How **large** is the **uterus**?

The uterus is normally 3 inches (7.5 centimeters) long and 2 inches (5 centimeters) wide. It weighs 1 to 1.4 ounces (30 to 40 grams). During pregnancy it can increase three to six times in size. At the end of gestation, the uterus is usually 12 inches (30 centimeters) long and weighs 2.4 pounds (1,100 grams). The total weight of the uterus and its contents (fetus and fluid) at the end of pregnancy is approximately 22 pounds (10 kilograms).

What are the **regions** of the **uterus**?

The uterus consists of the body and the cervix. The body is the largest area of the uterus. The superior part of the body is called the fundus. The cervix connects the uterus to the vagina.

Which medical **condition** is caused by the **abnormal location** of **endometrial tissue**?

Endometriosis is the abnormal location of uterine endometrial tissue. Under normal conditions, endometrial tissue only lines the uterus, but in some instances the tissue migrates to remote locations of the body such as the ovaries, pelvic peritoneum, the vagina, the bladder, or even the small intestine and lining of the chest cavity. Endometriosis is associated with dysmenorrhea (painful menstruation), pelvic pain, and infertility.

What is **pelvic inflammatory disease**?

Pelvic inflammatory disease (PID) is an infection of the uterus, uterine tubes, or other female reproductive organs. Many women do not experience any symptoms of PID while the disease is damaging their reproductive organs. When symptoms are present, pain in the lower abdominal area is most common. Other symptoms may include fever, unusual vaginal discharge that may have a foul odor, painful intercourse, painful urination, irregular menstrual bleeding, and pain in the right upper abdomen (rare).

During the 1920s George Papanicolaou (1883–1962) did research that showed a microscopic smear of vaginal fluid could detect the presence of cancer cells in the uterus. These findings were not generally accepted at the time by the medical community. Several years later, in 1943, he published *Diagnosis of Uterine Cancer by the Vaginal Smear* with Herbert F. Traut, a clinical gynecologist. This time, following publication of his findings, the medical community began to use the Pap smear as a diagnostic tool for cancer. The Pap smear is more than 90 percent reliable in detecting cancer, decreasing dramatically the mortality rate for cancer of the uterus and cervix.

What are the **complications** of **pelvic inflammatory disease**?

Permanent damage to the reproductive organs is a complication of pelvic inflammatory disease. Scar tissue often develops in the uterine tubes. Infertility is the result of the uterine tubes becoming blocked by scar tissue. As many as one in eight women becomes infertile due to PID. Ectopic pregnancies are more common in women who suffer from PID since a fertilized egg may remain in the uterine tubes and begin to grow. Finally, many women suffer from chronic pelvic pain.

Why is a **hysterectomy** recommended?

Hysterectomy is the surgical removal of the uterus. The most frequent reasons for a woman undergoing a hysterectomy are uterine fibroids, endometriosis, and uterine prolapse. Cancers of the pelvic organs account for only about 10 percent of all hysterectomies.

How does a **complete hysterectomy** differ from a **partial** or **radical hysterectomy**?

The most common type of hysterectomy is a complete or total hysterectomy. Both the cervix and uterus are removed in this procedure. A partial or subtotal hysterectomy removes only the upper part of the uterus and leaves the cervix in place. The most extensive hysterectomy is a radical hysterectomy which removes the uterus, the cervix, the upper part of the vagina, and supporting tissues. A radical hysterectomy is usually only performed in some cases of cancer.

What is the **female reproductive cycle**?

The female reproductive cycle is a general term to describe both the ovarian cycle and the uterine cycle, as well as the hormonal cycles that regulate them. The ovarian cycle is the monthly series of events that occur in the ovaries related to the maturation of an oocyte. The menstrual cycle is the monthly series of changes that occur in the uterus as it awaits a fertilized ovum.

What are the **phases** of the **reproductive cycle**?

The reproductive cycle consists of three phases: 1) the menstrual phase; 2) the preovulatory phase; and 3) the postovulatory phase. During each phase there are changes in the ovaries and in the uterus in response to hormonal secretions from the pituitary (follicle-stimulating hormone [FSH] and luteinizing hormone [LH]) and gonads (estrogens and progesterone).

Events of the Ovarian & Uterine Cycles During the Reproductive Cycle

	Ovarian Cycle	Uterine Cycle
Menstrual phase	20 or more secondary follicles begin to enlarge; follicular fluid accumulates in the follicle	Menstrual flow passes from uterus to exterior via vagina
Preovulatory phase	Follicular phase because follicles are growing and developing; secondary follicles continue to grow; one follicle outgrows the others and becomes the dominant follicle; estrogen secretion increases; secretion of FSH decreases	Proliferative phase because the endometrium is growing
Ovulation	Rupture of the mature follicle; release of the secondary oocyte into the pelvic cavity	
Postovulatory phase	Luteal phase; mature follicle collapses; if fertilization occurs, the corpus luteum continues past its normal two-week lifespan; if fertilization does not occur, the corpus luteum lasts for only two weeks and then degenerates; follicular growth then resumes and a new cycle begins	Preparation of endometrium for fertilized ovum; if fertilization does not occur, the menstrual phase begins again

Which event separates the **preovulatory phase** from the **postovulatory phase**?

Ovulation, the release of a secondary oocyte into the pelvic cavity, separates the preovulatory phase from the postovulatory phase.

How long the does female reproductive cycle last?

The reproductive cycle averages 28 days, but it may last from 24 to 35 days. A cycle of 20 to 45 days is still considered within the range of normalcy. The menstrual phase lasts five to seven days. The preovulatory phase lasts 8 to 11 days, and the postovulatory phase lasts an average of 14 days.

When does menarche occur?

Menarche (from the Greek, meaning "beginning the monthly") is the first menses, or menstrual cycle, of a girl's life. It usually begins between the ages of 11 and 12.

How does primary amenorrhea differ from secondary amenorrhea?

Amenorrhea is the absence of a menstrual period not due to pregnancy, breastfeeding, or menopause. Primary amenorrhea is diagnosed when girls have not had their first menstrual period by age 16. It may be caused by an endocrine, genetic, or developmental disorder. Secondary amenorrhea is diagnosed in previously menstruating women who have not had a menstrual period for six months or more. It may be caused by physical or emotional stress, including excessive weight loss, anorexia nervosa, depression, or grief. Many female athletes have secondary amenorrhea due to decreased levels of body fat.

How much blood is lost during menstruation?

Approximately 1.2 to 1.7 ounces (35 to 50 milliliters) of blood is lost during menstruation. In addition to blood, degenerating tissue cells from the endometrium of the uterus are also lost. They are replaced during the next uterine cycle.

What is menopause?

Menopause is the cessation of ovulation and menstrual periods. The supply of follicles in the ovaries is depleted, increasing the amount of follicle-stimulating hormone (FSH), while decreasing the amount of estrogen and progesterone. The process may take one to two years. The years preceding the final menstrual period are known as perimenopause.

What is the average age of when menopause occurs?

Menopause usually occurs between ages 45 and 55; the average age in the United States is 51 to 52 years.

What are the functions of the vagina?

The vagina is a muscular tube that extends between the uterus and the external genitalia. Its main functions are: 1) to serve as the passageway for the elimination of menstrual fluids; 2) to receive the penis during sexual intercourse and hold sperm prior to its passage into the uterus; and 3) to serve as the birth canal for the fetus during childbirth and delivery.

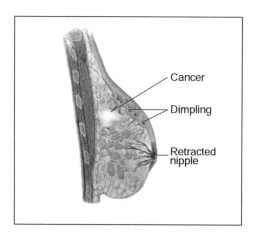

Signs of breast cancer. (From Bickley, L.S., and Szilagyi, P. *Bates' Guide to Physical Examination and History Taking.* 8th Ed. Philadelphia: Lippincott, Williams & Wilkins, 2003.)

How does the environment of the **vagina** protect it against **infections**?

The acidic environment of the vagina—its normal pH is 3.5 to 4.5—restricts the growth of many pathogenic organisms. If the acidity of the vagina is reduced, bacterial, fungal, or parasitic organisms may increase, thus increasing the risk of infections known as vaginitis. Yeast infections are common infections of the vagina.

Which organs comprise the **external female genitalia**?

The external female genitalia, collectively known as the vulva or pudendum, consist of the mons pubia, labia majora, labia minora, vestibular glands, clitoris, and vestibule of the vagina. The mons pubis (from the Latin, meaning "mountain" and "pubic") is a bulge of adipose tissue covered by pubic hair. The labia majora, two longitudinal folds of skin below the mons pubia, form the outer borders of the vulva. The labia minora lie between the labia majora. The labia majora and minora surround and protect the vaginal and urethral openings. The labia merge at the anterior point to form the clitoris. The clitoris becomes enlarged during sexual stimulation. The area between the labia minora is the vestibule. The vestibule contains glands and the openings of the vagina and urethra.

What are the **mammary glands**?

The mammary glands, located in the breasts, are modified sweat glands. They produce and secrete milk following childbirth in response to hormonal stimulation. Each breast has a pigmented projection, the nipple, surrounded by a darker pigmented area called the areola.

Where do most **breast cancers begin**?

Most breast cancers begin in the cells that line the ducts. The ducts are the passages that connect the lobules (the milk-producing glands) to the nipples. Some cancers begin in the cells of the lobules.

What **routine screening** tests are recommended to detect **breast cancer**?

Three screening tests may be recommended to detect breast cancer: 1) a screening mammogram, 2) clinical breast exam, and 3) monthly breast self-exam. The National Cancer Institute recommends that women in their forties have a screening mammogram every one to two years, unless they have certain risk factors that would require screening to begin at a younger age or more frequently.

Are all **lumps** found in the **breasts cancerous**?

Most lumps found in breast tissue are benign and not cancerous. Many of these lumps are fibrous, scarlike tissue and cysts (fluid-filled sacs).

SEXUAL RESPONSE AND CONCEPTION

What is the **human sexual response**?

The act of sexual intercourse, also called coitus (from the Latin *coire,* meaning "to come together"), involves a sequence of physiological and emotional changes in males and females. These changes are known as the human sexual response.

What are the **stages** of human **sexual response**?

Human sexual response may be divided into four stages: 1) excitement, 2) plateau, 3) orgasm, and 4) resolution.

Is the **sexual response** the same in **males** and **females**?

At one time it was believed that the sexual response in females was probably less intense and different than in males. Recent studies have shown that males and females experience similar feelings and sensations during sexual intercourse.

Which **reflexes** are **activated** during the excitement stage of the **human sexual response**?

During excitement, the initial stage of the human sexual response that is also called arousal, parasympathetic reflexes of the autonomic nervous system are activated. The parasympathetic impulses increase blood flow to the genitalia and secretion of lubricating fluids. In males, the penis becomes erect. In females, the clitoris and nipples become erect. These responses continue through the plateau stage, when the penis is inserted into the vagina.

What are the **responses** of males and females **during orgasm**?

Orgasm is marked by rhythmic contractions of the genital organs in both males and females. In males, ejaculation occurs during the orgasm stage of sexual response. During ejaculation sperm are released into the vagina and begin to swim to the uterus. Although in females there is no counterpart to ejaculation, rhythmic contractions of the vagina, uterus, and perineal muscles do occur.

Which **physiological changes** occur during **resolution**?

Breathing rate, heart rate, and muscles quickly return to the normal state during resolution. A feeling of relaxation usually accompanies resolution.

How many **sperm** are required to **ensure fertilization**?

Although millions of sperm may enter the vaginal canal during sexual intercourse, only one sperm enters and fertilizes an ovum. The additional sperm are necessary to increase the chance that one sperm will survive the acidic conditions of the vagina and will successfully complete the journey to the uterine tubes for fertilization. Once the sperm reaches the ovum it releases an enzyme that dissolves the outer wall of the ovum, allowing the sperm to enter. Following fertilization the outer membrane of the ovum thickens to prevent other sperm from entering the now fertilized egg.

What are some causes of **infertility**?

Infertility is defined as the inability to become pregnant after one year of trying. The cause of infertility may be with either the man or the woman or a combination of both partners. Female causes of infertility are often associated with problems with ovulation. The underlying reason may be premature ovarian failure (when the ovaries stop functioning prior to the onset of natural menopause); polycystic ovary syndrome (PCOS), when the ovaries do not regularly release an egg or do not release a viable egg; blocked fallopian tubes; physical problems with the uterine wall; or uterine fibroids. Male causes of infertility include the inability to have or sustain an erection, not having enough sperm or enough semen to carry the sperm to the egg, or having sperm that do not have the proper shape to move in the right way.

What is **assisted reproductive technology (ART)**?

According to the Centers for Disease Control (CDC), assisted reproductive technology includes all fertility treatments in which both the sperm and eggs are handled during the treatment. Most ART procedures involve surgically removing eggs from a woman's ovaries, combining them with sperm in the laboratory, and returning them to the woman's body or donating them to another woman.

How do the various **ART methods differ**?

One of the most successful and effective ART methods is in vitro fertilization (IVF). It may be used when the woman's fallopian tubes are blocked or when the man produces too few sperm. The woman takes a drug that causes the ovaries to produce multiple eggs. Once mature, the eggs are removed and put in a dish in the lab along with the man's sperm for fertilization. After three to five days, healthy embryos are implanted in the woman's uterus.

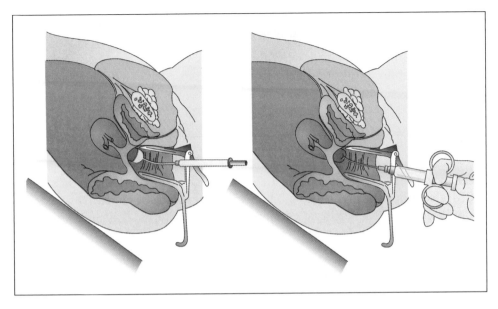

There are two different approaches to artificial insemination. Either the sperm is deposited next to the cervix (left) or injected into the uterine cavity (right). (From Pilliterri, Adele. *Maternal and Child Nursing*. 4th ed. Philadelphia: Lippincott, Williams & Wilkins, 2003).

Zygote intrafallopian transfer (ZIFT), also called Tubal Embryo Transfer, is similar to IVF. Fertilization occurs in the laboratory. Then the very young embryo is transferred to the fallopian tube instead of the uterus.

Gamete intrafallopian transfer (GIFT) involves transferring eggs and sperm into the woman's fallopian tube. Fertilization occurs in the woman's body. It is not as common a procedure as either IVF or ZIFT.

Couples in which there are serious problems with the sperm or who have been unsuccessful with IVF may try intracytoplasmic sperm injection (ICSI). In ICSI, a single sperm is injected into a mature egg. Then the embryo is transferred to the uterus or fallopian tube.

Are there **other fertility treatments** than assisted reproductive technology?

Fertility treatments may also include artificial (or intrauterine) insemination in which sperm (from the woman's husband, partner, or a donor) are injected into the woman's uterus, leading to conception. Women may also take medications to stimulate egg production. They may then be able to conceive without further medical intervention.

How is **pregnancy confirmed**?

Early external signs of pregnancy include a missed menstrual period, bleeding or spotting, fatigue, and tender, swollen breasts. A pregnancy test is needed to confirm

307

pregnancy. Pregnancy tests look for the hormone human chorionic gonadotropin (hCG), also called the pregnancy hormone, in either urine or blood. This hormone is produced when the fertilized egg implants in the uterus. It is only present in pregnant women.

Are **home pregnancy tests** more reliable than laboratory tests?

Home pregnancy tests check the urine for hCG. The amount of hCG increases daily in pregnant women, so most tests are fairly accurate about two weeks after ovulation. Laboratories may use one of two blood tests to confirm pregnancy. A qualitative hCG blood test checks to see whether or not the hormone is present. Its reliability is comparable to a urine test. A quantitative blood test (the beta hCG test) measures the exact amount of hCG in the blood. It can detect even tiny amounts of hCG, making it extremely accurate. Blood tests may confirm a pregnancy as early as six to eight days following ovulation.

What is the purpose of **contraceptive devices** and **methods**?

Contraceptive ("against conception") devices and methods aim to prevent pregnancy. The various contraceptive devices and methods use one of two techniques to avoid pregnancy: preventing sperm and ova from meeting or making the environment unsuitable for fertilization. Hormonal methods may either prevent the ovary from releasing an egg into the uterine tubes or cause changes in the cervix or uterus, making it difficult for sperm to enter the uterus or implant a fertilized egg in the uterus. Barrier methods, withdrawal, natural family planning, and sterilization prevent sperm and ova from meeting. Intrauterine devices interfere and prevent the implantation of a fertilized ovum in the uterus.

What are the main types of **birth control**?

Spermicides, condoms (male and female), the diaphragm, the cervical cap, and Lea's Shield are barrier methods. Oral contraceptives (birth control pills), injections of hormones, the vaginal ring, and skin patch are various types of hormonal methods of birth control.

Formerly called the "rhythm method," natural family planning is also now called periodic abstinence or fertility awareness. It involves a variety of different methods, including basal body temperature method, ovulation/cervical mucus method, symptothermal method, calendar method, and lactational amenorrhea to monitor the times of fertility during an individual woman's cycle.

Withdrawal requires the man to remove his penis from the woman's vagina prior to ejaculation thereby not allowing sperm to enter the woman's vagina. However, sperm may be present in the fluid prior to ejaculation, so this method is not very reliable.

Sterilization, called a vasectomy in men and tubal sterilization in women, permanently blocks the pathways for sperm and ova. In men, the vas deferens are cut so sperm cannot mix with semen. The tubes that carry sperm to the penis are clamped, cut, or sealed so that the ends do not heal and rejoin. In women, the fallopian tubes are closed by tying, banding, clipping, blocking, or cutting them, or by sealing them with electric current.

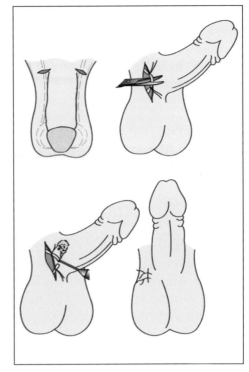

A vasectomy is an outpatient procedure that effectively sterilizes the patient. *LifeART image copyright © 2008. Lippincott, Williams & Wilkins.*

Is **surgical sterilization reversible**?

Sterilization, both male and female, should be considered a permanent, irreversible procedure. Tubal sterilization may be reversible, but it requires microsurgery to reconnect the uterine tubes after they have been blocked. Similarly, the reversal of a vasectomy in males is a lengthy, complicated operation requiring general anesthesia.

How are **sexually transmitted diseases (STDs)** transmitted?

Sexually transmitted diseases (STDs) are usually transmitted via sexual contact between individuals. Any form of sexual activity, including anal, oral, and vaginal sex, may spread an STD. In addition, sexually transmitted diseases may be spread from a mother to her baby during childbirth.

What are the some **common STDs**?

Sexually transmitted diseases may be caused by bacteria or viruses. Common STDs are chlamydia, genital warts (HPV), genital herpes (HSV), syphilis, gonorrhea, and chancroid. Viral hepatitis, especially hepatitis B, and AIDS are also considered STDs since they are frequently transmitted from one individual to another via sexual contact.

What is the **incidence of STDs** in the **United States**?

More than 15 million new cases of STDs are reported in the United States each year.

Incidence and Prevalence of Selected Sexually Transmitted Diseases

Sexually Transmitted Disease	Estimated New Cases per Year
Chlamydia	929,462
Herpes (HSV)	1,000,000
Genital Warts/Human Papillomavirus (HPV)	6,200,000
Gonorrhea	700,000
Syphilis	32,000
Trichomoniasis	7,400,000

What **complications** may arise from **STDs**?

Complications associated with STDs include pelvic inflammatory disease in women and epididymitis (inflammation of the epididymis) in men. Complications from STDs may cause infertility and increase the risk of some cancers. Blindness, bone deformities, and mental retardation may also be caused by STDs.

Which **STDs** most frequently cause **pelvic inflammatory disease**?

Pelvic inflammatory disease (PID) is most frequently caused by the bacteria found in chlamydia and gonorrhea.

HUMAN GROWTH AND DEVELOPMENT

INTRODUCTION

What is **growth** and **development**?

Growth is an increase in size. Human growth begins with a single fertilized ovum. As it grows, the number of cells increases as a result of mitosis, while at the same time the newly formed cells enlarge and grow in size. Development is the ongoing process of change in the anatomical structures and physiological functions from fertilization through each phase of life to death.

What are the **two divisions** of human **development**?

Human development is divided into the prenatal development phase and postnatal development phase. Prenatal development begins at the time of conception and continues until birth. Postnatal development begins at birth and continues to maturity, when the aging process begins and ultimately ends in death.

What is the **gestation period** of human development?

The gestation period is the time spent in prenatal development. Human gestation is conveniently divided into three trimesters of three months each. The first trimester is the period of embryological and early fetal development. During the second trimester the organs and organ systems develop. The third trimester is characterized by a period of rapid growth prior to birth.

How do the terms **zygote, embryo,** and **fetus** differ?

All three terms refer to the individual developing within the uterus of a woman following conception. Fertilization between sperm and ovum produces a zygote, a single cell consisting of 46 chromosomes. It is called a zygote during the first week of development. At the end of the first week of development, the zygote becomes an embryo. It is called an embryo from the second week of development through the eighth week of development. Beginning at the ninth week of development, it is called a fetus.

Why is **fetal development** described in terms of **lunar months** rather than **calendar months**?

Since the female reproductive cycle is closer to the lunar month of 28 days, fetal development is described in terms of lunar rather than calendar months. Each lunar month consists of four weeks for a total of 40 weeks, or ten lunar months.

How **large** is a **zygote** at the time of conception?

The single cell that is formed at conception, the zygote, is approximately 0.005 inches (0.135 millimeters) in diameter and weighs approximately 0.005 ounces (150 milligrams).

PRENATAL DEVELOPMENT–
EMBRYONIC PERIOD

How many **distinct stages** of development are part of the **prenatal period**?

The prenatal period of development consists of two distinct stages of development: embryological development and fetal development. Embryological development begins with fertilization and continues until the end of the eighth week of development. Fetal development begins at the ninth week of development and continues until birth.

What **four events** follow **fertilization**?

The four events that occur immediately following fertilization are cleavage, implantation, placentation, and embryogenesis. Immediately after fertilization, the single cell divides into two cells. During cleavage, these cells continue to divide. Each division brings two new cells called blastomeres. Each blastomere is approximately half the size of the parent cell. Cleavage occurs as the cells move from the uterine tube to the uterine cavity.

How long does the **journey** take from the **uterine tube** to the **uterine cavity**?

It takes about three days for the zygote to travel from the uterine tube to the uterine cavity. At the end of the journey, the solid mass of cells is called a morula (from the Latin *morum,* meaning "mulberry") because it resembles a mulberry.

When does the **blastocyst** form?

The morula hollows out into a fluid-filled sphere on day four or five following fertilization. It is now called a blastocyst. By the end of the first week, the blastocyst begins to implant in the uterus.

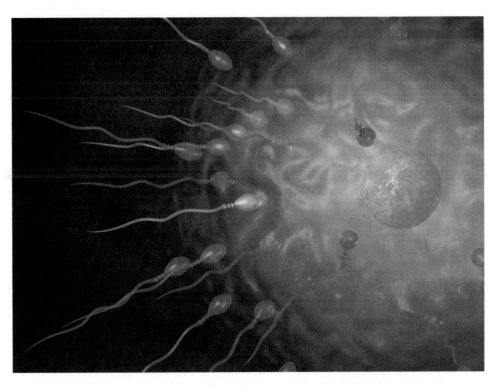

Swarms of sperm attempt to fertilize an egg at once, though typically only one sperm penetrates the ovum successfully. *LifeART image copyright © 2008, Lippincott, Williams & Wilkins.*

When does implantation occur?

Implantation, the attaching of the blastocyst to the endometrium of the uterus, begins about six to seven days following fertilization. By day nine, the blastocyst is completely enclosed by endometrial cells and is implanted in the uterus. Only after the blastocyst is implanted in the uterus can a pregnancy continue.

What is an ectopic pregnancy?

An ectopic pregnancy occurs when the fertilized ovum is implanted in an area other than the uterus. Most frequently, the fertilized ovum is implanted in the uterine tubes. A fetus in an ectopic pregnancy cannot survive because it does not receive nourishment from the uterus. An ectopic pregnancy must be terminated because it jeopardizes the health of the mother.

What are the functions of the placenta?

The placenta is a vascular structure that forms at the site of implantation between the cells that surround the embryo and the endometrium of the uterus. Its main functions are the exchange of gases, nutrients, and wastes between the maternal and fetal bloodstreams. In addition, the placenta secretes hormones. There is no actual blood flow between mother and fetus.

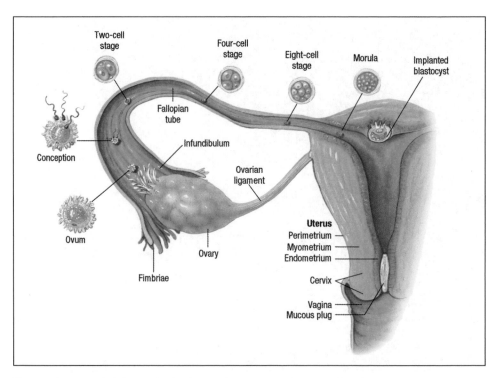

The egg is fertilized in the Fallopian tube, then begins cell division even before it is implanted in the uterine wall. *Anatomical Chart Co.*

Does the **placenta** continue to **grow** throughout a **pregnancy**?

The placenta grows rapidly until the fifth month of a pregnancy, at which time it is nearly fully developed. At the end of a pregnancy the placenta is about 1 inch (2.5 centimeters) thick and 8 inches (20.5 centimeters) in diameter. It weighs approximately 1 pound (0.45 kilograms).

How does the **umbilical cord form**?

The umbilical cord is formed from the extraembryonic or fetal membranes of the embryo during the fifth week of development. It contains two arteries and one vein. The arteries carry carbon dioxide and nitrogen wastes from the embryo to the placenta, and the vein carries oxygen and nutrients from the placenta to the embryo. The umbilical cord is usually 0.4 to 0.8 inches in diameter (1 to 2 centimeters) and 19 to 22 inches long (50 to 55 centimeters).

What is **embryogenesis**?

Embryogenesis is the process during which the embryo begins to separate from the embryonic disc. The body of the embryo and the internal organs begin to form at this point. The embryo becomes a distinct entity separate from the embryonic disc and the extraembryonic membranes. The left and right sides, as well as the dorsal and ventral surfaces, are now distinct.

Is there a scientific basis to save cord blood following delivery?

During the 1970s researchers discovered that umbilical cord blood was a source for blood-forming stem cells. Blood-forming stem cells are the early cells found primarily in the bone marrow that are capable of developing into red blood cells, white blood cells, and platelets. Certain serious illnesses, such as leukemia and lymphoma, are treated with bone marrow transplants to stimulate the growth of healthy blood cells. In situations in which there is a family history of diseases that may be treated with bone marrow transplants, parents may consider saving the cord blood for future use. However, the chances of a child developing a disease that will be treated with stem cell transplants are extremely low. In addition, there is little evidence to support that self-donated stem cells from cord blood are more successful than from a relative or other close match. Furthermore, the cost of collecting and storing cord blood for future use is expensive. The American Academy of Pediatrics does not recommend cord-blood banking for families who do not have a history of disease. Some parents may consider donating cord blood to a nonprofit cord blood bank for research or to save the life of another child.

How many **embryonic layers** are in the **embryonic disc**?

The embryonic disc contains three distinct embryonic layers: 1) the outer layer, called the ectoderm, which is exposed to the amniotic cavity; 2) the inner layer called the endoderm; and 3) the mesoderm. The mesoderm forms between the ectoderm and the endoderm.

Which **organs form** from the different **germ layers**?

All organs form from the three germ layers, as explained in the table below.

Organ and Organ System Formation from Germ Layers

Organ System	Ectodermal Layer	Mesodermal Layer	Endodermal Layer
Integumentary system	Epidermis, hair follicles and hairs, nails, sweat glands, mammary glands, and sebaceous glands	Dermis	
Skeletal system	Pharyngeal cartilages, portions of the sphenoid bone, the auditory ossicles, the styloid processes of the temporal bone, neural crest (formation of the skull)	All components except some pharyngeal derivatives	

315

Organ System	Ectodermal Layer	Mesodermal Layer	Endodermal Layer
Muscular system		All components	
Nervous system	All neural tissue, including brain and spinal cord		
Endocrine system	Pituitary gland and adrenal medullae	Adrenal cortex, endocrine tissues of heart, kidneys, and gonads	Thymus, thyroid gland, and pancreas
Cardiovascular system		All components	
Respiratory system	Mucous epithelium of nasal passageways		Respiratory epithelium and associated mucous glands
Lymphatic system		All components	
Digestive system	Mucous epithelium of mouth and anus, salivary glands		Mucous epithelium (except mouth and anus), exocrine glands (except salivary glands), liver, and pancreas
Urinary system		Kidneys, including the nephrons and the initial portions of the collecting system	Urinary bladder and distal portions of the duct system
Reproductive system		Gonads and the adjacent portions of the duct systems	Distal portions of the duct system, stem cells that produce gametes
Miscellaneous		Lining of the body cavities (pleural, pericardial, and peritoneal) and the connective tissues that support all organ systems	

What is the **amniotic cavity**?

The amniotic cavity is a fluid-filled chamber. It contains amniotic fluid that surrounds and cushions the developing embryo. At week 10 there is approximately 1 ounce (30 milliliters) of fluid in the cavity. Towards the end of a pregnancy, between weeks 34 and 36, there is about 1 quart (1 liter) of fluid.

What is the **temperature** of the **amniotic fluid**?

The temperature of the amniotic fluid is usually 99.7°F (37.6°C), slightly higher than the mother's body temperature.

When is **amniocentesis** usually performed and why?

Amniocentesis is usually done after the fifteenth week of pregnancy. It is a prenatal test used to screen for and identify genetic disorders or test for lung maturity. During the procedure, a thin needle is inserted into the amniotic cavity to remove some of the amniotic fluid, which contains fetal cells and various chemicals produced by the fetus.

What are some other **prenatal diagnostic techniques** used during pregnancy?

Ultrasonography, chorionic villi sampling, and alpha-fetoprotein screening are other prenatal diagnostic screening techniques besides amniocentesis. Fetal ultrasound is often done early in a pregnancy to determine whether it is an ectopic pregnancy. Fetal ultrasound is an accurate way to determine fetal age and predict a due date. Placental abnormalities, fetal growth and development (including heart rate), and congenital abnormalities may be detected with fetal ultrasound. It is a noninvasive test and relatively safe for the mother and fetus.

Chorionic villi sampling is another technique used to detect birth defects, such as Down syndrome or Tay-Sachs disease. It is usually done early in a pregnancy, between the ninth and fourteenth weeks. A sample of cells, called the chorionic villi, is taken from the placenta where it attaches to the wall of the uterus. The chorionic villi are tiny projections from the placenta that have the same genetic material as the fetus. The tissue sample is taken either through the cervix or through the abdominal wall.

The alpha-fetoprotein (AFP) test is a screening test to determine whether a woman is at risk for carrying a fetus with birth defects. AFP is produced by the fetus and appears in the mother's blood during a pregnancy. Abnormally high amounts of this protein may indicate a problem with the fetus. It is usually done between weeks 16 to 18 in the pregnancy.

When do **significant changes** occur during the **early stages** of **prenatal development**?

Several significant changes occur during the first two weeks of prenatal development, as shown in the accompanying table.

Significant Changes in the First Two Weeks of Prenatal Development

Time Period	Developmental Stage
12–24 hours following ovulation	Fertilized ovum
30 hours to third day	Cleavage
Third to fourth day	Morula (solid ball of cells is formed)
Fifth day through second week	Blastocyst
End of second week	Gastrula (germ layers form)

What are some major **developmental events** during the **embryonic period**?

At the end of the embryonic period (eighth week of development), all of the major external features (ears, eyes, mouth, upper and lower limbs, fingers, and toes) are formed and the major organ systems are nearing completion.

Major Developmental Events during the Embryonic Period

Time Period	Major Developments
Week three	Neural tube, primitive body cavities and cardiovascular system form
Week four	Heart is beating; upper limb buds and primitive ears visible; lower limb buds and primitive eye lenses appear shortly after ears
Week five	Brain develops rapidly; head grows disproportionately; hand plates develop
Week six	Limb buds differentiate noticeably; retinal pigment accentuates eyes
Week seven	Limbs differentiate rapidly
Week eight	Embryo appears human; external ears are visible; fingers, toes lengthen; external genitalia are visible, but are not distinctly male or female

How large is the **embryo** at the **end** of the embryonic period?

The embryo is about 0.75 inches (19 millimeters) in length at the end of the embryonic period (end of the eighth week).

How large is the **fetus** at the end of the **first trimester**?

At the end of the first three months of pregnancy, the fetus is nearly 3 inches (7.6 centimeters) long and weighs about 0.8 ounces (23 grams).

PRENATAL DEVELOPMENT– FETAL STAGE

What is the **purpose** of the **fetal stage** of development?

During the fetal stage, beginning after the eighth week of development, the fetus increases in size. All organs formed during the embryonic stage mature to the point where they can function at birth.

What are some major **developmental events** during the **second trimester** of pregnancy?

The second trimester of a pregnancy lasts from weeks 13 through 27. Each week brings changes and new developments in the fetus.

Major Developmental Events during the Second Trimester

Time Period	Major Developments
Week 13	Baby begins to move, although the movements are too weak to be felt by the mother; ossification of bones begins
Week 14	Prostate gland develops in boys; ovaries move from the abdomen to the pelvis in girls
Week 15	Skin and hair (including eyebrows and eyelashes) begins to form; bone and marrow continues to develop; eyes and ears are nearly in their final location
Week 16	Facial muscles are developing allowing for facial expressions; hands can form a fist; eggs are forming in the ovaries in girls
Week 17	Brown fat tissue begins to develop under the skin
Week 18	Fetus is able to hear such things as the mother's heartbeat
Week 19	Lanugo and vernix cover the skin; fetal movement is usually felt by the mother
Week 20	Skin is thickening and developing layers; fetus has eyebrows, hair on the scalp and well-developed limbs; fetus often assumes the fetal position of head bent and curved spine
Week 21	Bone marrow begins making blood cells
Week 22	Taste buds begin to form; brain and nerve endings can process the sensation of touch; testes begin to descend from the abdomen in boys; uterus and ovaries (with the lifetime supply of eggs) are in place in girls
Week 23	Skin becomes less transparent; fat production increases; lungs begin to produce surfactant, which will allow air sacs to inflate; may begin to practice breathing
Week 24	Footprints and fingerprints begin to form; inner ear is developed, controlling balance
Week 25	Hands are developed, although the nerve connections are not yet fully developed
Week 26	Eyes are developed; eyebrows and eyelashes are well-formed; hair on head becomes fuller and longer
Week 27	Lungs, liver, and immune system are developing

What is the purpose of **vernix** and **lanugo**?

Vernix is a white, pasty, cheese-like coating on the skin consisting of fatty secretions from the sebaceous glands and dead epidermal cells. It protects the skin of the developing fetus. Lanugo is a very fine, silk-like or down-like hair that covers the skin. It may help to hold the vernix on the skin.

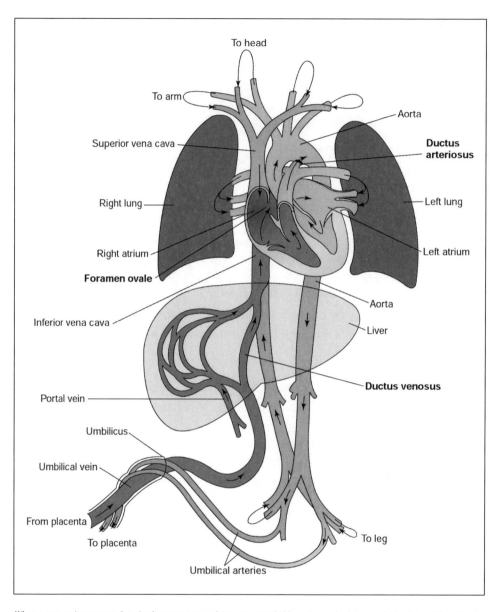

To head

To arm

Superior vena cava

Right lung

Right atrium

Foramen ovale

Inferior vena cava

Portal vein

Umbilicus

Umbilical vein

From placenta

To placenta

Umbilical arteries

Aorta

Ductus arteriosus

Left lung

Left atrium

Aorta

Liver

Ductus venosus

To leg

When a woman is pregnant, her circulatory system undergoes a remarkable metamorphosis to supply the fetus with nutrients. (From Pilliterri, Adele. *Maternal and Child Nursing*. 4th ed. Philadelphia: Lippincott, Williams & Wilkins, 2003).

How does **blood circulate** in the **fetus**?

Fetal circulation differs from circulation after birth because the lungs of the fetus are nonfunctional. Therefore, blood circulation essentially bypasses the lungs in the fetus. The umbilical vein carries oxygenated blood from the placenta to the fetus. About half of the blood from the umbilical vein enters the liver, while the rest of the blood bypasses the liver and enters the ductus venosus. The ductus venosus joins the inferior vena cava. Blood enters the right atrium of the heart and then flows through the foramen ovale to the left atrium. Blood then passes into the left ventricle (lower portion of the heart) and then to the aorta. From the aorta, blood is sent to the head and upper extremities. It returns to the right atrium of the heart through the superior vena cava. Some blood stays in the pulmonary trunk to reach the developing lung tissues.

How does **fetal blood** differ form **adult blood**?

Fetal blood has a greater oxygen-carrying capacity than adult blood. Fetal hemoglobin can carry 20 to 30 percent more oxygen than adult hemoglobin.

What is a normal **fetal heart rate**?

The fetal heart rate is much faster than an adult's (or even a child's) heart rate. The average resting heart rate is 60 to 80 beats per minute. The normal fetal heart rate is 110 to 160 beats per minute.

How does the **fetal circulatory system change** at birth?

Immediately after birth, an infant no longer relies on maternal blood to supply oxygen and nutrients. As soon as the baby begins to breathe air, blood is sent to the lungs to be oxygenated. The ductus arteriosus, the special fetal vessel connecting the aorta and pulmonary valve, is no longer needed and closes. A separate left pulmonary artery and aorta form after birth. In addition, the foramen ovale, the special opening between the left and right atria in the heart, closes and normal circulation begins.

What are some major **developmental events** during the **third trimester** of pregnancy?

During the third trimester of a pregnancy the fetus continues to grow, while the organ systems continue to develop to the point of being fully functional. Fetal movements become stronger and more frequent.

Major Developmental Events during the Third Trimester

Time Period	Major Developments
Week 28	Eyes begin to open and close; fetus has wake and sleep cycles
Week 29	Bones are fully developed, but still pliable; fetus begins to store iron, calcium, and phosphorus

Time Period	Major Developments
Week 30	Rate of weight gain increases to 0.5 pounds (227 grams) per week; fetus practices breathing; hiccups are not uncommon
Week 31	Testes begin to descend into scrotum in boys; lungs continue to mature
Week 32	Lanugo begins to fall off
Week 33	Pupils in the eyes constrict, dilate, and detect light; lungs are nearly completely developed
Week 34	Vernix becomes thicker; lanugo has almost disappeared
Week 35	Fetus stores fat all over the body; weight gain continues
Week 36	Sucking muscles are developed
Week 37	Fat continues to accumulate
Week 38	Brain and nervous system are ready for birth
Week 39	Placenta continues to supply nutrients and antibodies to fight infection
Week 40	Fetus is fully developed and ready for birth

How much does the **fetus grow** during each month of pregnancy?

During the early weeks of development there are great changes from the embryonic to the fetal stages, but the overall size of the embryo is very small. As the pregnancy continues, weight gain and overall size becomes much more significant. Until the 20th week of pregnancy, length measurements are from the crown (or top) of the head to the rump. After the 20th week, the fetus is less curled up, and measurements are from the head to the toes.

Average Size of the Fetus during Pregnancy

Gestational Age (weeks)	Size	Weight
8	0.63 in (1.6 cm)	0.04 oz (1 g)
12	2.13 in (5.4 cm)	0.49 oz (14 g)
16	4.57 in (11.6 cm)	3.53 oz (100 g)
20	6.46 in (16.4 cm)	10.58 oz (300 g)
24	11.81 in (30 cm)	1.32 lb (600 g)
28	14.8 in (37.6 cm)	2.22 lb (1 kg)
32	16.69 in (42.4 cm)	3.75 lb (1.7 kg)
36	18.66 in (47.5 cm)	5.78 lb (2.62 kg)
40	20.16 in (51.2 cm)	7.63 lb (3.46 kg)

When is the **fetus** considered **full-term**?

The fetus is considered to be full-term at the end of the 37th week of pregnancy.

What are the **complications** of **premature birth**?

Babies born before the 37th week of gestation (preterm babies) are very small and fragile. Birth weights are often less than two pounds. Many of the organ systems are

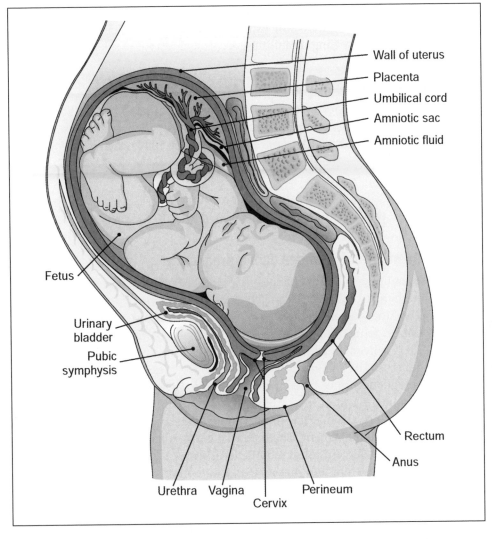

A fetus late in its third trimester has almost become fully formed. (Cohen, B. J., and Wood, D. L. *Memmler's The Human Body in Health and Disease*. 9th Ed. Philadelphia: Lippincott, Williams & Wilkins, 2000.)

not fully developed, which leads to complications as these infants struggle to survive. Complications include:

- Inability to breathe or breathe regularly on their own due to underdeveloped lungs
- Difficulty in body temperature regulation; the baby cannot maintain his or her own body heat
- Feeding and growth problems because of an immature digestive system
- Jaundice due to a buildup of bilirubin
- Anemia due to not enough red blood cells to carry oxygen to tissues
- Bleeding into the brain

Although after a year or two most preterm babies are developmentally the same as full-term babies, some may still experience breathing difficulties, hearing or vision problems, and learning disabilities.

How **common** are **premature births**?

According to the Centers for Disease Control, the preterm birth rate for 2005 in the United States was 12.7 percent. The percentage of infants delivered at less than 37 completed weeks of gestation has risen 20 percent since 1990, when it was only 10.6 percent.

BIRTH AND LACTATION

What are the **maternal changes** during **pregnancy**?

There are several physiological changes in a mother during pregnancy, in addition to changes in the size of the uterus and changes in the mammary glands. The mother must eat, breathe, and eliminate wastes for both herself and her developing fetus, which is totally dependent upon the mother. The mother's respiratory rate goes up so her lungs can deliver the extra oxygen and remove the excess carbon dioxide generated by the fetus. The maternal blood volume increases by nearly 50 percent by the end of a pregnancy, since blood flowing into the placenta reduces the volume of blood throughout the rest of the cardiovascular system. Because the mother must also nourish the fetus, she may feel hungry more often, and her nutritional requirements increase 10 to 30 percent. The maternal glomerular filtration rate increases by approximately 50 percent to excrete the fetus's waste. Consequently, the combination of increased weight and pressure on the mother's urinary bladder and the elimination of additional waste products lead to more frequent urination.

How do the **mammary glands develop** during **pregnancy**?

The mammary glands increase in size in response to placental hormones and maternal endocrine hormones. The areola darken in color. Clear secretions, such as colostrum, are stored in the duct system of the mammary glands. The secretions may be expressed from the nipple.

What are the **stages of labor**?

The goal of labor is the birth of a new baby. Labor is divided into three stages: 1) dilation, 2) expulsion, and 3) placental. Delivery of the fetus occurs during expulsion.

How is the **onset of labor** identified?

Different women will experience different symptoms at the onset of labor. Some women may experience lower back pain or cramping similar to menstrual cramps. In some women, the amniotic sac ruptures early in labor with a sensation of fluid leak-

Engagement, Descent, Flexion

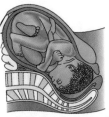

Internal Rotation

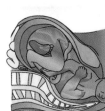

External Rotation (Restitution)

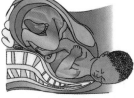

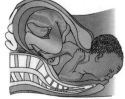

Extension Beginning (rotation complete)

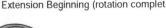

External Rotation (Shoulder rotation)

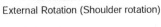

Extension Complete

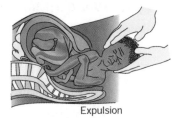

Expulsion

In a normal birth, the baby's head emerges first and may be positioned either sideways (right) or in a more vertical, twisted position (left). (From Pillitteri, A. *Maternal and Child Nursing.* 4th Ed. Philadelphia: Lippincott, Williams & Wilkins, 2003.)

ing either as a trickle or a large gush of fluid. Some women will lose the mucous plug that blocks the cervix with a brownish or red-tinged mucous discharge. Ultimately, as labor progresses, uterine contractions become more powerful and more frequent.

What events occur during **dilation**?

The purpose of dilation is to dilate (open) and thin (efface) the cervix to permit the fetus to move from the uterus into the vagina. Dilation is divided into three phases: 1) the early labor phase, 2) the active labor phase, and 3) transition. During the early phase, contractions last 30 to 60 seconds and occur every 5 to 20 minutes at regular intervals. As labor progresses, the frequency of the contractions will increase. The cervix dilates from 0 to 3 centimeters during the early phase of labor. During active labor, contractions become stronger, last longer (45 to 60 seconds or longer) and occur at more frequent intervals (as frequent as every 2 to 4 minutes). The cervix dilates from 3 to 7 centimeters. The final phase of dilation is transition. During transition, the cervix dilates from 7 to 10 centimeters. It is now fully dilated. Contractions during transition last 60 to 90 seconds with sometimes not even a minute between contractions.

What is the **placental stage** of labor?

During the placental stage of labor, uterine contractions separate the connections between the endometrium and the placenta. The placenta, along with the fetal membranes and any remaining uterine fluid, are ejected through the birth canal.

How does **false labor** differ from true labor?

Uterine contractions that are neither regular nor persistent are false labor. Oftentimes in false labor the contractions may stop when the mother walks or even shifts positions. The contractions in true labor become stronger, more frequent, and do not cease. Once true labor begins it will continue until the fetus is delivered.

How **long** does **labor** last?

The length of labor differs with every woman. Even the same woman will experience labor differently with each pregnancy. In general, dilation is the longest stage of labor. It can last for several hours to several days, especially for the first-time mother. The early phase of labor is the longest. Active labor may last from three to eight hours, although it can be shorter or longer. Transition is the shortest part of dilation. It may last for only 15 minutes. Expulsion (delivery) may take only a few minutes to several hours. Delivery of the placenta usually takes only 5 to 10 minutes and is usually no longer than 30 minutes.

What is the difference between **fraternal** and **identical twins**?

Fraternal, or dizygotic twins, develop when a woman ovulates two separate oocytes that are fertilized by two different sperm. Fraternal twins do not resemble each

> ## How do healthcare providers induce labor?
>
> **W**hen labor does not begin on its own, healthcare providers may induce labor. The most common method of inducing labor is to give the hormone oxytocin. Oxytocin will start contractions and keep them strong and regular. Labor is generally induced when the pregnancy has lasted two weeks beyond the due date or if there is a concern the baby will be too large for safe delivery. Labor may also be induced if the health of the mother becomes endangered.

other any more than other brothers and sisters of the same parents resemble each other. They may be of the same sex or different sexes.

Identical, or monozygotic twins, develop from the same fertilized ovum. The blastomeres may separate early in cleavage, or the inner cell mass may split prior to gastrulation. Identical twins have the same genetic makeup because they are formed from the same pair of gametes. They look alike and are the same sex.

Do **twins share** a **placenta, umbilical cord,** or **amniotic sac**?

Identical twins will often share the same placenta, but usually have separate amniotic sacs. Each twin always has its own, separate umbilical cord. Nonidentical twins have separate placentas, amniotic sacs, and umbilical cords.

How does a **multiple pregnancy** affect the **mother's health**?

Multiple pregnancies (pregnancies with more than one fetus) pose special risks since the strains on the mother (e.g., the need for oxygen and other nutrients for each fetus) are multiplied. Preeclampsia (high blood pressure and protein in the urine) and gestational diabetes are more common in multiple pregnancies.

What are some of the **risks and complications** associated with **multiple pregnancies**?

One of the most common risks of a multiple pregnancy is preterm birth. On average, most twin pregnancies last 35 weeks, while pregnancies with triplets last only 33 weeks. Pregnancies with quadruplets last only 29 weeks on average. Low-birth weight (less than 5.5 pounds, or 2.5 kilograms) due to preterm birth or poor fetal development is also very common in multiple pregnancies. Babies weighing less than 3.34 pounds (1.5 kilograms) at birth are at a greater risk for lasting disabilities, including mental retardation, cerebral palsy, and hearing and vision loss. Lung problems and breathing difficulties are common in babies born before 34 weeks.

How **common** are **multiple births**?

In 2004, 132,219 twins were born in the United States. This represented 3.2 percent of all live births (a birth ratio of 32 per 1,000 births). Only 0.2 percent of live births were triplets or higher-order multiple births, representing a birth ratio of 176.9 per 100,000 live births.

Why has the number of **multiple births increased** over the last decade?

The ratio of multiple births has increased 32 percent since 1994. Reasons for the increase include mothers having babies after age 30 and using fertility drugs and other methods of assisted reproduction.

What are **conjoined twins**?

Conjoined twins are identical twins whose embryonic discs do not separate completely. They typically share some skin and an organ, often the liver, and perhaps other internal organs. If the fusion is minor, they may be separated surgically with relative ease. On rare occasions, conjoined twins are joined at the head or share so many organs that it is nearly impossible to separate them.

When does **milk** begin to be **produced** in a pregnant woman?

By the end of the sixth month of a pregnancy, the mammary glands are developed in order to secrete milk. They begin to secrete colostrum. Once the placenta is delivered and the secretion of estrogen and progesterone drop, milk production increases.

What are the **stimuli** for **milk release**?

The hormones prolactin and oxytocin are involved in milk production and release. An infant's sucking stimulates the release of these two hormones.

What is the **composition** of human **breast milk**?

Human breast milk consists of mostly of water (88 percent), sugars (6.5 to 8 percent), lipids (3 to 5 percent), proteins (1 to 2 percent), amino acids, and salts. It also contains large quantities of lysozymes—enzymes with antibiotic properties.

Human milk is bluish-white in color and sweet. The blue color comes from the protein and the white comes from the fat. There are approximately 750 calories per liter of breast milk.

How does **colostrum** differ from breast milk?

Colostrum is the first fluid secreted by the mother's breasts in the first several days after delivery. It is higher in proteins and has less fat than milk. It also contains a high concentration of antibodies that protect the baby from infections until his own immune system matures. A mother produces approximately 3 ounces (100 ccs) of colostrum in a 24-hour period.

How much **milk** does a mother **produce**?

A nursing mother produces 850 to 1,000 milliliters of milk each day. Mothers of multiples will naturally produce enough milk for each infant.

What are the **benefits** of **breastfeeding**?

Breastfeeding provides benefits to both the baby and the mother. A major benefit to the baby is that breast milk supplies the correct amount of nutrients as the baby grows from an infant to a healthy toddler. The nutrients in breast milk also protect the infant from certain childhood illnesses. Finally, recent research has shown that breast milk contains certain fatty acids (building blocks) that help the infant's brain develop.

In the early days following childbirth, the mother's body releases a hormone that makes her uterus contract and get smaller in response to the baby's sucking. Breastfeeding also provides many emotional benefits between mother and child and encourages maternal-infant bonding.

POSTNATAL DEVELOPMENT

What are the **stages** of **postnatal development**?

The five life stages of postnatal development are: 1) neonatal, 2) infancy, 3) childhood, 4) adolescence, and 5) maturity. The neonatal period extends from birth to one month. Infancy begins at one month and continues to two years of age. Childhood begins at two years of age and lasts until adolescence. Adolescence begins at around 12 or 13 years of age and ends with the beginning of adulthood. Adulthood, or maturity, includes the years between ages 18 to 25 and old age. The process of aging is called senescence.

What are some **developmental changes** that occur during the **neonatal period**?

The greatest change from birth to the neonatal period is that the neonate must begin to perform many functions that had previously been done by the mother, especially respiration, digestion, and excretion. With the first breath of air following delivery, the lungs fill with air and the neonate begins to breathe for himself.

How do **heart rate** and the rate of **respiration** differ between a **neonate** and an **adult**?

The average neonate heart rate is 120 to 140 beats per minute, compared to a resting heart rate of 60 to 80 beats per minute in an adult. The average respiratory rate for a neonate is 30 breaths per minute, compared to 12 to 28 breaths per minute in an adult.

What are the major **developmental milestones** during **infancy**?

A normal infant will double his or her birth weight by five or six months of age and triple his or her birth weight during the first year of life. Major developmental milestones during infancy are summarized in the table below. There is considerable variation between individuals, but these are within the normal range.

Major Developmental Milestones During Infancy

Age	Major Milestones in Average Infant
End of first month	Bring hands to face; move head from side to side while lying on stomach; hear very well and often recognize parents' voices
End of third month	Raise head and chest while lying on stomach; open and shut hands; brings hands to mouth; smile; recognize familiar objects and people
End of seventh month	Roll over stomach to back and back to stomach; sit up; reach for objects with hand; support whole weight on legs when supported and held up; enjoy playing peek-a-boo; begin to babble
End of first year	Sit up without assistance; get into the hands and knees position; crawl; walk while holding on; some babies are able to take a few steps without support; use the pincer grasp; use simple gestures— e.g., nodding head, waving bye-bye
End of second year	Walk alone; begin to run; walk up and down stairs; pull a toy behind them; say single words (15–18 months); use simple phrases and two-word sentences (18–24 months); scribble with a crayon; build a tower with blocks

Does **growth** continue at a **constant pace**?

Growth is most rapid during the prenatal period. During infancy and childhood growth slows until the time of puberty.

What is the average age when **puberty begins**?

The average age when puberty begins in the United States today is around 12 years in boys and 11 years in girls. The normal range is 10 to 15 years in boys and 9 to 14 years in girls.

What **hormonal events** signal the onset of **puberty**?

Three different hormonal events occur that signal the onset of puberty (from the Latin *puber,* meaning "adult"). The hypothalamus increases production of gonadotropin-releasing hormone (GnRH). This stimulates the endocrine cells in the anterior lobe of the pituitary gland, causing circulating levels of follicle-stimulating hormone (FSH) and luteinizing hormone (LH) to rise rapidly. Finally, in response to increased levels of FSH and LH, the ovaries and testes secrete increased amounts of androgens and estrogens. The secondary sex characteristics appear, gamete production begins, and there is a sudden increase in the growth rate, culminating in the closure of the epiphyseal cartilages.

What are the major **body changes** at **puberty**?

In addition to general body changes that occur in both males and females at puberty, physical changes in the genitalia, skin, hair growth, and voice are collectively termed the secondary sex characteristics.

Secondary Sex Characteristic Changes in Males and Females

Area of Body	Males	Females
General body changes	Shoulders broaden, muscles thicken and height increases; body odor from armpits and genitals becomes apparent; skeletal growth ceases by about age 21	Pelvis widens; fat distribution increases in hips, buttocks, breasts; skeletal growth ceases by about age 18
External genital organs	Penis increases in size; scrotum enlarges; penis and scrotum become more pigmented	Breasts enlarge; vagina enlarges and vaginal walls thicken
Internal genital organs	Testes enlarge; sperm production increases in testes; seminal vesicles, prostate gland, bulbourethral gland enlarge and begin to secrete	Uterus enlarges; ovaries secrete estrogens; ova in ovaries begin to mature; menstruation begins
Skin	Secretions of sebaceous gland thicken and increase, often causing acne; skin thickens	Estrogen secretions keep sebaceous secretions fluid, inhibit development of acne and blackheads
Hair growth	Hair appears on face, pubic area, armpits, chest, around anus; general body hair increases; hairline recedes in the lateral frontal regions	Hair appears on pubic area, armpits; scalp hair increases with childhood hairline retained
Voice	Voice becomes deeper as larynx enlarges and vocal cords become longer and thicker	Voice remains relatively high pitched as larynx grows only slightly

How does **adolescence** differ from **puberty**?

Adolescence (from the Latin *adolescere,* meaning "to grow up") begins at puberty and ends at adulthood when physical growth stops. Puberty is the point when an individual becomes physiologically capable of reproduction. Adulthood begins between ages 18 and 25.

How is **senescence** defined?

Senescence (from the Latin *senex,* meaning "old") is the process of aging. Physiological changes continue to occur even after complete physical growth is attained at maturity. As people age, the body is less able to and less efficient in adapting to environmental changes. Maintaining homeostasis becomes harder and harder, especially when the body is under stress. Ultimately, death occurs when the combination of stresses cannot be overcome by the body's existing homeostatic mechanisms.

What are some general **effects of aging** on the human body?

The aging process affects every organ system. Some changes begin as early as ages 30 to 40. The aging process becomes more rapid between ages 55 and 60.

Effects of Aging

Organ System	Effect of Aging
Integumentary	Loss of elasticity in the skin tissue, producing wrinkles and sagging skin; oil glands and sweat glands decrease their activity, causing dry skin; hair thins
Skeletal	Decline in the rate of bone deposition, causing weak and brittle bones; decrease in height
Muscular	Muscles begin to weaken; muscle reflexes become slower
Nervous	Brain size and weight decreases; fewer cortical neurons; rate of neurotransmitter production declines; short-term memory may be impaired; intellectual capabilities remain constant unless disturbed by a stroke; reaction times are slower
Sensory	Eyesight is impaired with most people becoming far-sighted; hearing, smell, and taste are reduced
Endocrine	Reduction in the production of circulating hormones; thyroid becomes smaller; production of insulin is reduced
Cardiovascular	Pumping efficiency of the heart is reduced; blood pressure is usually higher; reduction in peripheral blood flow; arteries tend to become more narrow
Lymphatic	Reduced sensitivity and responsiveness of the immune system; increased chances of infection and/or cancer
Respiratory	Breathing capacity and lung capacity are reduced due to less elasticity of the lungs; air sacs in lungs are replaced by fibrous tissue
Digestive	Decreased peristalsis and muscle tone; stomach produces less hydrochloric acid; intestines produce fewer digestive enzymes; intestinal walls are less able to absorb nutrients

| Excretory | Glomerular filtration rate is reduced; decreased peristalsis and muscle tone; weakened muscle tone often leads to incontinence |
| Reproductive | Ovaries decrease in weight and begin to atrophy in women; reproductive capabilities cease with menopause in women; sperm count decreases in men |

Glossary

Absorption—The passage of substances, such as nutrients, across membranes into cells or body fluids.

Acid—A substance that releases hydrogen ions when dissolved in water. Acidic solutions have a pH below 7.0.

Acid-base balance—A system that regulates and maintains the pH of blood at approximately 7.35-7.45.

AIDS—Acquired Immune Deficiency Syndrome [CB1](also known as Acquired Immunodeficiency Syndrome) is a disease caused by the human immunodeficiency virus (HIV). It is characterized by a deficiency of helper T cells, which results in the severe impairment of the immune system.

Allergy—A reaction to an antigen; also called hypersensitivity.

Amino acid—The structural unit of a protein; an organic chemical compound consisting of an amino group ($-NH_2$) and a carboxyl group ($-COOH$).

Anabolism—The metabolic process in which smaller molecules form more complex, larger molecules.

Anatomy—The branch of science that studies the structure of the body.

Anterior—Toward the front of the body; also called ventral.

Antibiotic—A substance used to destroy pathogens.

Antibody—A protein produced by the immune system in response to an antigen, which is then destroyed by the body.

Antigen—A substance that induces the production of antibodies.

Appendicular—Pertaining to the upper or lower limbs.

Artery—A blood vessel that transports blood away from the heart to the capillary beds.

ATP—Adenosine triphosphate is an organic molecule consisting of adenine, ribose, and three phosphate groups. It stores and releases energy for chemical reactions in cells.

Atrophy—A wasting away or decrease in size of tissue, such as muscle fibers, usually from lack of use.

Autoimmune disorder—Any disorder in which there is an immune response to normal cells and tissues in which the body forms antibodies to its own antigens.

Autonomic neuron—A nerve cell that innervates or stimulates smooth muscle cells and glands.

Axial—Pertaining to the head, neck, and trunk of the body.

Axon—Tubular extensions of a neuron that transmit nerve impulses away from the cell body, often to another neuron.

Bacteria—Single-celled microorganisms common in the environment and on the body; some are pathogens.

Base—A substance that releases hydroxyl ions when dissolved in water. Basic solutions have a pH greater than 7.0.

Benign—Not malignant or otherwise a threat to life.

Biofeedback—A technique to control or manipulate bodily responses that are typically involuntary.

Blood—A loose connective tissue whose matrix is plasma and that contains red blood cells, white blood cells, and platelets. Blood circulates throughout the body, carrying nutrients and oxygen to all parts of the body and carrying away waste products.

Bone—A rigid connective tissue that has a matrix of collagen fibers embedded in calcium salts, bone is the hardest tissue in the body.

Buffer—A substance that stabilizes and neutralizes the pH of a solution by releasing or removing hydrogen ions.

Cancer—A malignancy characterized by uncontrolled growth and replication of affected cells, cancer is a disease that spreads locally by invasion of surrounding tissue and systemically by metastasis.

Carcinogen—A cancer-causing agent.

Cartilage—A connective tissue with an abundant number of collagen fibers in a rubbery matrix.

Catabolism—The metabolic process that breaks down complex organic molecules into simpler components, catabolism is accompanied by the release of energy.

Cell—The smallest structural and functional unit of life.

Chemoreceptor—A sensory receptor, such as taste buds, that responds to chemical stimuli.

Cholesterol—A lipid that maintains the strength and flexibility of cell membranes; also, the molecule from which steroid hormones and bile acids are synthesized.

Clone—A group of cells that originate from a single cell and are therefore identical genetically to each other and the parent cell.

Connective tissue—One of the basic types of tissue consisting of fibers and widely spaced cells in an extracellular material called the matrix, connective tissue includes bone, blood, cartilage, collagen, adipose, and loose connective tissue.

Cutaneous—Pertaining to the skin.

Dendrite—Thin, highly branched extensions of neurons that receive signals.

Digestion—The breakdown of large molecules (e.g., ingested materials), chemically and mechanically, into smaller, simpler molecules that can be absorbed by the cells of the digestive tract.

Distal—Away from the trunk or midline of the body.

DNA—Deoxyribonucleic acid is a double-stranded nucleic acid containing sugar, deoxyribose, a nitrogenous base, and a phosphate group. DNA contains genetic information.

Dorsal—Toward the back; also called posterior.

Double helix—The structural arrangement of DNA consisting of two strands of polynucleotides.

Embryo—The prenatal stage of development beginning with implantation and ending at the end of the eighth week of development.

Epidermis—The outer epithelial layer of skin.

Epithelial tissue—A type of tissue consisting of groups of cells which form a superficial covering or internal lining of a vessel or cavity.

Fetus—The prenatal stage of development beginning at the ninth week of development and ending at birth.

Forensics—Applying scientific knowledge to legal problems; forensic medicine applies medical facts to legal problems.

Gamete—A sex cell (ovum or egg cell in a female, sperm cell in a male) that contains half the normal number of chromosomes.

Ganglia—Groups of neuron cell bodies outside the central nervous system.

Gastrointestinal tract—The digestive tract that begins at the mouth and ends at the anus. The upper gastrointestinal tract consists of the oral cavity, esophagus, and stomach. The lower gastrointestinal tract consists of the small and large intestine (colon) and ends at the anus.

Gene—A specific sequence of DNA that contains the molecular recipe for a subunit of a protein called a polypeptide. A gene carries hereditary information.

Genetics—The study of heredity and inherited variations.

Genome—The complete set of genes inherited by offspring from their parents.

337

Gland—Groups of specialized cells that produce secretions.

Histology—The study of tissues.

Homeostasis—The state of inner balance and stability maintained by the human body despite constant changes in the external environment.

Hormone—A chemical compound secreted by endocrine glands that circulate via the bloodstream to affect the metabolic activities of cells in another part of the body called "target" cells.

Immunity—Resistance to foreign substances as a protective mechanism.

In vitro—Outside the living body and in an artificial environment (e.g., in vitro fertilization taking place in a test tube).

In vivo—Within the living body.

Infection—The invasion and establishment of pathogens in body tissues.

Inferior—Directional term meaning below.

Keratin—The tough fibrous protein portion of the epidermis, hair, and nails.

Lateral—Pertaining to the side.

Lymph—The fluid of the lymphatic system that is similar to plasma but does not contain erythrocytes (red blood cells) or platelets.

Medial—Toward the midline of the body.

Meiosis—Cell division that divides the genetic material (chromosomes) in half, resulting in gametes (egg and sperm cells).

Membrane—A thin sheet or layer consisting of epithelium and the underlying connective tissue; membranes line body cavities and cover or separate regions, structures, and organs.

Metabolism—The total chemical and physical processes occurring in the body at a given time; may be anabolic or catabolic.

Metastasis—The spread of cancer cells from the primary site of the disease to another part of the body, establishing secondary tumors in another part of the body.

Mitosis—A step of cell division in which a single cell nucleus divides to produce two identical daughter cell nuclei.

Motor neuron—A nerve cell that innervates and stimulates muscle cells.

Muscle—An organ comprised of muscle cells and fibers, blood vessels, nerves, and connective tissue that contracts and relaxes to move body parts.

Nerve—A bundle of nerve fibers held together by layers of connective tissue.

Neuron—Nerve cell; specialized cells of the nervous system that consist of a cell body with a nucleus, dendrites, and axons that produce impulses or nerve signals.

Neurotoxin—A poisonous substance that affects the nervous system.

Neurotransmitter—A chemical substance released by one neuron that affects another neuron; neurotransmitters form the basis of communication between neurons and may produce an inhibitory or excitatory response.

Nutrients—A chemical compound that can be broken down to supply the body with energy and nourishment.

Organ—A group of different tissues working together to perform a specific function.

Organ system—A group of organs working together to perform a specific function.

Pathogen—A disease-causing agent.

Pathology—The branch of science that studies diseases, including their symptoms and causes.

pH—The measurement of the concentration of hydrogen ions (H^+) in an aqueous solution. The pH scale, ranging from 0 to 14, with 7 being neutral, is used to measure the acidity or alkalinity of the solution. The higher the number from 7 to 14, the more basic (alkaline) the solution; the lower the number, from 0 to 7, the more acidic the solution.

Physiology—The study of the functions of the body's parts and organs.

Plasma—The liquid portion of blood.

Platelet—Fragments of cytoplasm manufactured in the bone marrow that play an important role in helping blood clot; also called thrombocyte.

Posterior—Toward the back; also called dorsal.

Proximal—Toward the trunk or midline of the body.

Pulmonary—Pertaining to the lungs.

Reflex arc—A series of events involving a sensory neuron, motor neuron, and sometimes an interneuron that lead to a reflex.

Renal—Pertaining to or involving the kidney.

Respiration—The exchange of gases, especially oxygen and carbon dioxide, between cells and the environment.

RNA—Ribonucleic acid is a nucleic acid consisting of ribose sugar found in the nucleus and cytoplasm of the cell associated with cellular activities.

Schwann cell—A type of cell in the peripheral nervous system that forms myelin.

Sensory—Relating to the senses or sensation.

Sphincter—A circular muscle that contracts to close the opening of a tubular structure.

STD—Sexually transmitted disease; a disease or infection that is transmitted from one individual to another via sexual contact.

Steroid—An organic molecule that includes four connected rings of carbon atoms, such as cholesterol.

Subcutaneous—Beneath the skin.

Superior—A directional term meaning above; toward the top of the body.

Synapse—The site of intercellular communication where a nerve impulse passes from one neuron to another.

Tissue—A group of similar, specialized cells that perform a specific function.

Tumor—A mass of tissue formed by the abnormal growth and rapid replication of cells; tumors can be benign or malignant.

Ulcer—A break in the skin or mucous membrane with the loss of the surface tissue; a lesion.

Vaccine—A preparation containing antigens that is administered to elicit an immune response to a disease.

Valve—A structure that briefly closes a passage or opening, allowing movement of fluid in one direction only.

Vein—A blood vessel that carries blood towards the heart.

Ventral—Toward the front of the body; also called anterior.

Virus—An infection-causing, protein-coated fragment of DNA or RNA genetic material, viruses are incapable of living on their own and only reproduce within their host cell.

Vital signs—Signs of life in a person, including pulse rate, respiratory rate, and body temperature. Blood pressure is often included as a vital sign.

Zygote—A fertilized ovum (egg cell) formed by joining a male and female gamete.

Index

Note: (ill.) indicates photos and illustrations.

A

"ABCD" rule, 49
Abdomen, 10
Abdominal reflex, 124
Abdominopelvic cavity, 11
ABO system, 195
Absorption, 251
Absorption of nutrients, 259
Accessory glands, 56–57, 296.
  See also Hair;
  Integumentary system;
  Nails; Skin
    cutaneous glands, 56–57
    mammary glands, 57
    sweat glands, 56–57
    whitehead vs. blackhead, 56
Accessory organs, 265–70,
  287–88. *See also* Digestive
  system; Kidneys; Lower
  gastrointestinal tract;
  Metabolism and nutrition;
  Upper gastrointestinal tract;
  Urinary system; Urine
    alcohol and liver, 268
    bile, 267
    bilirubin, 267
    cirrhosis of the liver, 268
    digestive juice, 267
    enzymes, 265, 267
    gall bladder, 265, 269–70
    gallstones, 270
    hepatitis, 268–69
    liver, 265, 267
    liver cancer, 268
    nonviral hepatitis, 268
    pancreas, 265, 267
    pancreatic enzymes, 267
    pregnant women and uri-
      nation, 287

small intestine, 267
ureter, 287
urethra, 288
urinary bladder, 287, 288
  (ill.)
urinary tract infections,
  288
urine, 287
viral hepatis, 268
Acetylcholine, 100, 110, 111
Achilles heel/tendon, 97
Acidity, 19
Acquired immunity, 221
Acquired immunodeficiency
  syndrome (AIDS), 223, 225
Acromegaly, 169
Action potential, 108–9, 110
Active immunity, 221
Adam's apple, 238
Addison's disease, 176
Adduction, 90
Adenoid, 215
Adenosine diphosphate
  (ADP), 26, 102
Adenosine triphosphate
  (ATP), 26–27, 27 (ill.)
    aerobic metabolism, 103
    creatine phosphate, 87
    muscle metabolization,
      102, 104
    oxygen, 18
Adipose tissue, 31, 33
Adolescence, 329, 332
ADP, 26, 102
Adrenal cortex, 175
Adrenal glands, 175–77,
  281–82. *See also*
  Hormones; Pancreas;
  Parathyroid glands; Pineal
  gland; Pituitary gland;

Reproductive organs;
Thyroid gland
  Addison's disease, 176
  adrenal cortex, 175
  adrenal medulla, 177
  corticosteroids, 175
  Cushing's syndrome,
    176–77
  glucocorticoids, 176–77
  hormones, 175, 177
  mineralocorticoid hor-
    mone, 176
  physical characteristics,
    175
Adrenal medulla, 177
Adrenocorticotropic hormone
  (ACTH), 166
Adulthood, 329
Aerobic metabolism, 103
Aerobic training, 86–87
Afferent neurons, 106
Age spots, 43–44
Aging, 104, 144, 246–47,
  332–33
Agranulocyte, 192
AIDS, 223, 225
Air movement, 247
Albinism, 43, 44 (ill.)
Albumin, 194
Alcohol, 268
Alkalinity, 19
Allergens, 230
Allergic reaction, 228–31, 229
  (ill.)
Allergies, 228–31. *See also*
  Lymphatic system;
  Lymphatic vessels and
  organs; Nonspecific
  defenses; Specific defenses
    anaphylactic shock, 230

343

Ilium, 75
Imaging techniques, 11–15
  blood flow, 13
  CAT scans, 11–12, 14
  mammography, 15
  nuclear magnetic reso-
    nance (NMR), 13
  nuclear magnetic reso-
    nance imaging, 13–15
  obstetrics, 15
  positron emission tomog-
    raphy (PET imaging),
    12–13
  ultrasound, 15
  X-rays, 11, 12, 12 (ill.), 13,
    14
Immediate allergic reactions,
  229
Immune system, 7, 51, 212–13
Immunity, 221
Immunodeficiency diseases,
  212–13
Immunoglobulin E, 228
Immunoglobulins, 219–21
Immunology, 209
Immunosuppressants, 228
Implantation, 312, 313
In vitro fertilization, 306, 308
Inactivated (killed) vaccine,
  227
Incisors, 254
Incontinence, 290
Incus, 148
Indole, 264
Inducing labor, 327
Infancy, 329, 330
Infection, 39, 219, 222, 304
Inferior (directional plane), 8,
  9 (ill.)
Inferior trunk, 10
Inferior vena cava, 204
Infertility, 301, 306
Inflammation, 36, 38, 218
Inflammatory bowel diseases,
  264
Infrasonic sound, 151
Ingestion, 251
Inhibitory neurotransmitters,
  111
Inhibitory postsynaptic
  potential, 109
Innate immunity, 221
Inner endocardium, 198
Insertion, 96
Insomnia, 139
Insulin, 178–79
Integrators, 39

Integumentary system, 7,
  41–57. *See also* Accessory
  glands; Hair; Nails; Skin
  effects of aging, 332
  organs, 41, 315
Intelligence, 116, 134
Intelligence measurement,
  133 (ill.), 133–34
Intelligence quotient (IQ),
  134
Interferons, 218
Intermediate hairs, 52
Internal respiration, 244
Interneurons, 34–35
Interstitial spaces, 213
Intestines, 183. *See also*
  Large intestine; Small
  intestine
Intramembranous
  ossification, 66
Involuntary muscle
  movements, 86
Iodized salt, 172
Ion channels, 108
IQ, 134
Iris, 154
Iron, 18
Iron lung, 242
Irregular bones, 63–64
Irritable bowel syndrome,
  264
Ischemic stroke, 120
Ischium, 75
Islet of Langerhans, 177–78
Itakura, Keiichi, 180

J

Jarvik, Robert K., 200
Jejunum, 260, 261
Jenner, Edward, 226
Jet lag, 181
Joints, 77–82. *See also*
  Appendicular skeleton; Axial
  skeleton; Bones; Skeletal
  system
  arthritis, 81
  artificial, 82
  classification of, 77
  dislocation, 80
  double-jointed, 81
  fibrous, 77, 79
  functional classes, 77, 79
  knee, 81
  knuckles, 80
  osteoarthritis, 82
  pregnancy, 79

pubic, 79
  rheumatoid arthritis, 82
  structural classes, 77, 79
  synovial, 79–81
  synovial fluid, 79
  types of, 78 (ill.)
Journal of Physiology, 4

K

Keratin, 42, 50
Keratinized cells, 50
Keratinocytes, 42, 51, 54
Kernig's reflex, 125
Kidney failure, 284, 286
Kidney stone, 284
Kidneys, 280–87. *See also*
  Accessory organs; Urinary
  system; Urine
  adrenal glands, 281–82
  aging, 284
  blood filtration, 284
  chronic renal disease, 286
  dialysis, 285 (ill.), 286
  failure, 284, 286
  fluid removal from blood,
    284
  hormones, 183
  kidney stone, 284
  location of, 280
  major vessels, 283–84
  nephron, 282, 283 (ill.)
  parts of, 281, 281 (ill.)
  protection of, 281
  red blood cell production,
    284
  size of, 280
  surviving with one kidney,
    282
  transplants, 287
  urea, 283
  vasa recta, 283
  Vitamin D, 282
Kissing, 88
Knee joint, 81
Knee-jerk reflex arc, 124
Knuckles, 80
Köhler, Georges, 222
Korotkoff, Nikolai, 207
Kyphosis, 72

L

Labor, 324–26
Labyrinth, 150
Lachrymal gland, 158
Lactose intolerance, 21–22

351